IMPACT

IMPACT

How Rocks from Space

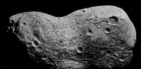

Led to Life, Culture, and Donkey Kong

GREG BRENNECKA

WILLIAM MORROW

An Imprint of HarperCollinsPublishers

HarperCollins books may be purchased for educational, business, or sales promotional use. For information, please email the Special Markets Department at SPsales@harpercollins.com.

FIRST EDITION

Title page art by Nazarii/Adobe Stock

Library of Congress Cataloging-in-Publication Data has been applied for.

ISBN 978-0-06-307892-5

22 23 24 25 26   LSC   10 9 8 7 6 5 4 3 2 1

To Celeste, Haloumi, and Cosmo. And to those brave enough to search for answers, be themselves, and enjoy life.

Contents

IMPACT

Introduction

AND SO IT BEGINS

The Universe. The Solar System. Earth. Life. Humanity. Religion. And of course, Donkey Kong. These are all immense and fascinating subjects in their own right, but there is something they all have in common, something physical that connects these diverse dots together. If you were to poll a million random people on what that connection is between those dots, probably zero would pick meteorites—but the reality is that rocks flying around the cosmos not only built our physical world and laid the foundation for life to exist, they also have had an inordinate influence on the various nonconcrete constructions of civilization. Meteorites are not just museum relics or interesting items to buy on the Internet as physical reminders of the death of the dinosaurs; they represent the origins of Earth and humanity.

Rocks from outer space not only create stories, but they tell them also. These scientific objects link the formation of the Solar System to the present day, acting as time capsules of information that extend billions of years. Humanity's pursuit of knowledge about the creation and evolution of the physical environments throughout the Universe utilizes meteorites in a variety of ways, as these rocks are often the only windows we have into the environments in which they were created.

Meteorites may be cool trinkets to some, they may be terrifyingly deadly objects to others, they can indeed be functional doorstops, but meteorites are also incredible scientific tools for studying the past. This book is a discussion of how meteorites have influenced our planet—since it was created until the present day—and what we have learned about our physical environments once we started to study meteorites as scientific objects.

The Beginning of Time and Space

When you are looking through a telescope, either in your backyard or at images from something as spectacular as the Hubble Space Telescope, you are essentially looking back in time. If you focus your telescope on something a million light-years away, then the light you are perceiving is a million years behind what actually is happening in that part of the Universe. On human time scales, this would be akin to just now being able to watch the 2015 World Series for the first time. Because the Universe has been expanding for such a long time, there are some objects that are pretty darn far away from us, both in space and time, and this allows astronomers to look at innumerable objects billions and billions of years in the past, informing us how galaxies and stellar systems form and evolve. If we look *really* far back to the start of it all—the "Big Bang" 13.7 billion years ago—we have to use a combination of telescopes, particle physics, and loads of mathematics. When we do this, it is very clear that the Universe was a very different place than it is now. First of all, there were only a few elements: hydrogen and helium were the only major players, and only traces of lithium and beryllium were to be found. That was it—no aluminum, no iron, no neon, and no einsteinium. For quite some time, it was essentially just an expanding hot cloud of protons, neutrons, electrons, and probably cockroaches. After a while, nuclear fusion started inside early stars; nuclear fusion is basically a way of merging ingredients like hydrogen and helium to get heavier elements—a process that ratchets up over time, creating even heavier elements. And since stars started forming, they also eventually started dying.* When a star reaches the end of its life, one way or another, it spews its guts over the cosmos, providing seed materials (elements like iron or neon) for the next generation of stars to repeat the process with a slightly heavier starting point. This process is what Carl Sagan was referring to on an episode of PBS's much-loved *Cosmos*:

> *The cosmos is also within us. We're made of star stuff.*

This famous line may seem an oft-repeated trope to some, but regardless of whether you have heard this many times before or if this is the first, this quote captures an incredible amount. It is simple, but it gets at the heart of what the Universe is: a giant recycling program.

The primary reason we know so much about this impressive recycling program is our ability to look backward in time at other star systems using telescopes, but, unfortunately for those interested more locally on how our own stellar system formed and evolved, we are restricted to present-day programming only. We know that our Solar System was born from the remnants of exploded stars: the Sun, the planets, comets, meteorites, Jimmy Fallon, etc. are all products of previous generations of stellar systems; this is just the latest iteration of the assemblage of those particular atomic particles. But how do we learn about the early days of the planetary bodies familiar to us? The light we get from the Sun only takes 8 minutes and 20 seconds to get to Earth, so the looking-back-in-time trick is not particularly helpful if we are trying to study something that happened in the Solar System over 4.5 billion years ago. Luckily for us, there remains a way to look back in time at our stellar neighborhood, and that is in the form of fossils that recorded the important events of our Solar System's origins.

---

* The lifetime of a star is primarily determined by its size. The larger the star, the shorter its lifetime. Even though a large star has a *lot* of fuel, it uses it up very fast. Smaller stars don't burn as bright or as hot as large stars, but they last a lot longer.

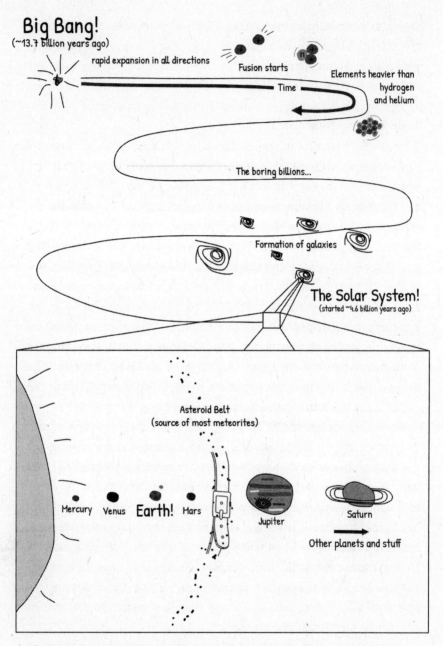

Big Bang!
(~13.7 billion years ago)

rapid expansion in all directions

Fusion starts

Time

Elements heavier than hydrogen and helium

The boring billions...

Formation of galaxies

The Solar System!
(started ~4.6 billion years ago)

Asteroid Belt
(source of most meteorites)

Mercury   Venus   Earth!   Mars   Jupiter   Saturn

Other planets and stuff

Categorically not to scale, and only a partially accurate timeline of the Universe and accompanying snapshot of the modern Solar System.

These fossils are meteorites, and these physical remains are largely the reason we know so much about the formation and evolution of our pettily little Solar System.

So What Is a Meteorite?

Meteorites go by a lot of names depending on their stage of existence. They are asteroids when orbiting the Sun, meteors for a literal hot-second as they streak across the sky, and they become meteorites once they land as an alien interloper on Earth. But they are all the same objects, just different names for different times. Think of meteorites like the musician Prince. He was born in Minnesota as Prince Rogers Nelson, but became an international superstar known simply as "Prince." He then changed his name to the unsayable symbol of love, at which time he was widely known as "The artist formerly known as Prince." He eventually dropped the symbol moniker to go back to his given first name. Yet, for all this name changing, he never strayed from being the same eclectic and talented individual dressed primarily in purple. This is essentially how it is with asteroids/meteors/meteorites, minus the killer guitar solos and flamboyant stage presence.

With the nomenclature stuff out of the way, very simply stated, most meteorites are pieces of our early Solar System that never really matured into legitimate planets. They are gatherings of cosmic dust and galactic garbage that coalesced into a stony object that then, fortuitously for us, found its way onto the surface of Earth. Most meteorites are thought to come from the asteroid belt, a loosely organized collection of floating rocks between Mars and Jupiter. The main asteroid belt is a mix of many different things: dust, small pebbles, large primitive chunks of Solar System material, broken pieces of former planets, and stuff that was never big enough to call anything other than the dismissive-sounding "planetesimal."

The cosmic debris that is now the asteroid belt is held in a relatively stable orbit due to the gravitational dance mostly between Jupiter and the Sun,* but occasionally objects of the asteroid belt run into one another. When these collisions happen, bits can get knocked off the offending parties and out of orbital bliss. If any chunk of the Solar System falls out of stable orbit, the object will career toward the biggest gravity well in the area, normally the Sun. Since Earth is located between the asteroid belt and the Sun, it can occasionally get in the way, and you guessed it, mass extinction. Or, much more likely, a souvenir-size space rock; it really depends a lot on the dimensions at hand.

If it is not an extinction-level-size rock, this can be a really great thing for us humans. Such chunks of the past have been floating around doing nothing but holding information about the past, largely unchanged, for an incredible expanse of time. This is billions of years with *nothing* happening. This incredible expanse of time, and the relative lack of things going on in space, means that meteorites give us snapshots into what was going on in the neighborhood when these rocks formed so long, long ago. Basically, this is just another way to look back in time, only with physical samples instead of telescopes.

The Scientific Study of Meteorites

The Meteoritical Society, the sole international society engaged in the study and dissemination of information about meteorites, has a membership of fewer than one thousand people. And while this number has been growing steadily over the decades, this includes retirees, part-time researchers, and graduate students. As such, the number of active, full-time professional meteorite researchers is probably more around one hundred, and this is for the entire world. That is not a lot of people, and, as with any issue, it is always fun to look to Florida for a bit of perspective. The Sunshine State has somewhere around two hundred people who make a living exclusively from alligator husbandry. Think about that for just a bit. Meteorites have extinguished hundreds of apex

---

* Saturn and the other gas giants play a role, but as with Sumo wrestling, influence scales with mass and location.

predator species virtually instantaneously and plunged the entire world into extended periods of darkness multiple times. It could conceivably be our misfortune as human beings to be snuffed out by a large space rock. Yet there are more people that have chosen to farm alligators in one state in the United States than currently study meteorites in the entire world. And I am not particularly advocating against the existence of exotic reptile products, just simply pointing out that we have more people actively farming an animal that most of the human population is terrified of than are studying samples of other planets that most of the human population is reportedly fascinated by. If that is also not serious motivation for a popular science book on meteorites, I don't know what is.

I like to think of the study of meteorites as "cosmic forensics." Meteorites were witnesses to a crime (formation of the Solar System), and people that study meteorites interrogate them to give up their information. And since our interrogation techniques routinely include slicing meteorites into small pieces, dissolving various bits with acids, and shooting interesting parts with high-powered lasers, meteoriticists sound a bit like James Bond villains. But we interrogate meteorites in the name of science; we are far less interested in world domination. At least most of us.

However grotesque it may sound, the study of space rocks searches for answers to a few basic, overarching questions. Ask anyone in the field why they do what they do and they will likely say they want to answer some variant of some or all these few questions:

- How did our Solar System form?
- What sequence of events occurred for the Solar System to turn out the way it did?
- Are there other systems out there similar to ours?
- How/Why did life develop in the Solar System?
- How unique is the development of life in the Universe?

By any measure, these are big questions, and likely questions many inquisitive people on the planet have. And these questions are answerable—at least approachable, but only through space science, in which meteorites play an outsize role. Meteorites provide the starting composition of what our Sun and planets are built from. Meteorites contain clocks that recorded how long certain processes took; meteorites contain thermometers and other fun things to look at the environmental conditions when the Sun was just getting started. Meteorites represent the "ground truth" for the models that dynamists make about how stellar systems form and evolve. Meteorites contain diamonds older than the Sun, pieces of long-exploded stellar systems, and some even shockingly contain amino acids and large amounts of water, the building blocks of life as we know it.

Reasons to Study Meteorites

Aside from the fact that knowing stuff is just cool and looking for answers that have not yet been found is an exciting and noble pursuit, there are many practical reasons to study extraterrestrial objects. For example, we are running out of palladium, a precious metal we need to make more smartphones. Most of the palladium we mine arrived by meteorite, because the original palladium from Earth's formation is all in the core. Since, as humans, we are not really that patient of a bunch and are not okay with just waiting for more metal-laden meteorites to arrive to save us from our palladium addiction, multiple companies have been formed to investigate mining asteroids for raw materials.

Maybe you are less of the consumeristic type and think more about the survival of the human species. After all, it is inevitable that the Earth will eventually become uninhabitable, either by our own destruction or from outside forces. If you are looking to understand more about why Earth is habitable, knowing the ingredients and conditions of the Solar System so you can look for other habitable places is a good place to start, and that requires meteorites. After all, they contain a record of essentially the entire history of the Solar System, including all the starting materials we had to work with. Obviously, finding Earth 2.0 will also include awesome telescopes, a really fast spacecraft, and some chic, shiny spacesuits, but it might be smart to know what you are looking for before you strap yourself into that supercool rocket ship in your form-fitted Versace self-breather spacesuit.

And while these practical reasons are great, for me at least, and I would guess many of my colleagues, studying meteorites as scientific objects is primarily for the joy of finding out information that was previously unknown. It is stimulating, it is exciting, and it is far less dangerous than farming alligators.

Beyond the overwhelming scientific information meteorites have provided, the path that meteorites traveled to be viewed as scientific objects is a fascinating historical account that spans the majority of human existence. Human-meteorite interactions occurred prior to humans having the capability to document anything, and meteorites played important roles in the cradles of civilization: ancient Mesopotamia, Egypt, and China all have fascinating meteoritic interactions that influenced how these cultures developed. The broad story of how meteorites affected cultural trajectory includes many fascinating twists, a U-turn or two, some larger-than-life personalities, the emergence of some of the world's most followed religions, gruesome emperor assassinations, people eating space rocks to become godlike, and even notes on how to build a proper house. Accounts of human interactions with meteorites over history are strange, sometimes humorous, oftentimes frustrating, but always interesting.

After reading this work, my hope is that you will agree that meteorites are far more than just rocks from space that occasionally kill things—they are incredibly important objects that played a crucial role in building our planet and our culture.

One

Important Early Meteorite Strikes

We tend to think of the Solar System as a fairly predictable place where the planets faithfully twirl around the Sun in ordained orbits, and where moons do a similar, but smaller-scale waltz with their planetary pals. Occasionally, maybe a comet from the outer reaches of the Sun's pull whizzes by for a bit of fiery fun every now and again, but overall, thanks to gravity, cosmic order is dependably maintained. And when you consider only the blink of time that humans have been on Earth, this homeostatic view is not too far off from reality. Contrary to this relative serenity, however, the early Solar System was a dynamic beast of chaos: formation and destruction of planetary-size bodies, intense irradiation, and general mayhem were the norm. But after the first few tens of millions of years of planetary pandemonium, things slowed down a bit. This perhaps still should not be called "stable" when, at any moment, Earth could be hit by a rogue asteroid the size of Mount Everest, flash-melting continent-size volumes of rock, resulting in an ocean of magma, but at least those types of things don't happen *that* often.

The Early Early Days of the Solar System

One important thing to remember about the Solar System is that it did not used to look like it does now. First, the Sun's behavior was very different from what it is today. When a young stellar object like the Sun first forms, it passes through a variety of phases before it settles in. For our star, the Sun, you can think of this as the "terrible ~2-millions." Even though the Sun was fainter at its inception than it is today, it was far more violent. This is partly due to the fact it was rotating much faster, essentially giving it a bit more energy. This faster rotation and extra energy resulted in higher magnetic fields, stronger solar winds, and orders of magnitude stronger ultraviolet and X-ray emissions from what we have today. Of course, it is perfectly cromulent to ask how we know anything about the young Sun, since, well, it was a long time ago and we were not there. Let's jump into the fun realm of astrophysics for a brief moment. The properties of a star, such as its luminosity and stellar lifetime, greatly depend on the star's size.* We know the size of the Sun and therefore the amount of nuclear fuel it can burn (mostly hydrogen and helium), so all we have to do is simply integrate the Lagrangian congruency to the point where it diverges from its inverse dielectric constant . . . carry the one . . . math, math, math . . . words, words, words . . . and bingo, history of the Sun revealed! Perhaps another good way is to just look at the gazillions of other stars out there in the Universe and see how those of a similar size are reacting during the different stages of their life cycles. Both of these methods lead us to the same basic conclusion: the young Sun was a temperamental and difficult child to be around.

Whereas the activity of the young Sun was not making things easy, there were also other, possibly even more catastrophic hazards to worry about for planets growing up in our young Solar System. That is right, planetary bullying . . . or at least smashing into one another while jockeying for positions that constituted stable orbits. The early

---

* You are likely now expecting some sort of "size matters" quip, but that is too predictable, and I will not oblige with anything more than this footnote.

Solar System was considerably more crowded with planetesimals and rocky objects than it is today, and importantly, the modern orbits of many of the planets have evolved since the Solar System started.* This combination of migrating planets and lots of stuff floating around led to a lot of planetary rock fights that did not end pleasantly for most of the parties involved. In particular, the large masses and corresponding gravitational power of the giant planets like Jupiter and Saturn caused these two planets to play an outsize role in defining the structure of the current planetary orbits. Because it took these gas giants a couple hundred million years to finally settle into their respective homes in the disk, their inward and outward migrations massively upset the order of things, relentlessly flinging large rocky bodies around crashing into one another—or more likely the Sun—until everyone finally found their happy place.

After the bulk of the migratory mess finished, the result was four rocky planets (Mercury, Venus, Earth, Mars) and the asteroid belt inside the orbit of the four gas giants (Jupiter, Saturn, Uranus, Neptune). Interestingly, based on the currently available planetary surveys from other stellar systems, our planetary arrangement is certainly not a common one. In the vast majority of exoplanet systems, gas giants tend to be close to the central star and not so distal, like in the Solar System. It is possible that due to the difficulty in detecting smaller rocky planets compared to gas giants, we presently have a skewed view of extrasolar planetary systems. However, at this point the general difference between the planetary arrangement in the Solar System and those in other parts of the galaxy appears to be real and may suggest that our setup is special. Is this arrangement important for why Earth is the only known planet that fosters life? We need to find a lot more stellar systems with rocky planets, and a few more planets that host life before we can answer any of these types of interesting questions. So, let's do that as soon as possible.

----

* This is commonly referred to as the "Grand Tack" model. This is because the scale of the tack, or diversion in planetary course, is grand.

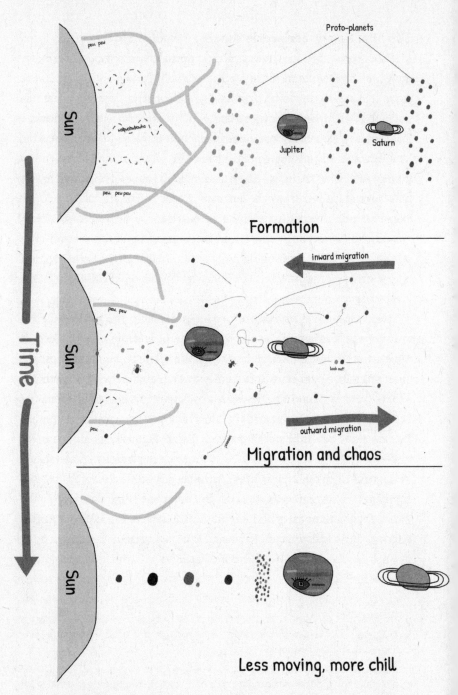

The intense activity of the young Sun and the migration of giant planets caused some serious havoc back in the day.

The First Known Meteorite Strike on Earth

It is a fairly well-established fact that we have a moon. On most evenings, it is a prominent fixture in the night sky, and well, we have been there a couple times to hit golf balls and drop a flag or two. What was not realized for the vast majority of human history was why or how we have a moon, especially such a large one.* In large part due to the samples returned during the Apollo missions and the armies of scientists that have studied them, we have an answer to that question, and as you may have surmised from the primary subject of this book, the Moon exists because of a really, really big meteorite.

Long before humans started having interesting and culturally consequential relationships with meteorites, space rocks have been assaulting Earth, basically since it was formed. The surface of the Earth is pitted with hundreds of craters from meteorite strikes dating back billions of years, all of which likely had interesting consequences not only for the local geology and biology, but for the entire evolution of the planet. However, unequivocally, the most consequential of these ancient meteorite impacts was the one that happened when the Earth was a mere toddler, less than ~150 million years after the birth of the Solar System. There are no craters from this impact because it was far too large to leave any such obvious scars. This is because the impact flash-melted the entire surface of the Earth and large portions of its mantle. The impactor itself, a Mars size body that has been named Theia,† was completely obliterated as it violently introduced itself to a fledgling Earth. During their meeting, significant portions of the silicate components (mantle and crust) of Earth and Theia were ejected off the surface

---

* Of all the "real" planets (sorry, Pluto) with moons, the Earth-Moon system is by far the closest in size to one another.

† This name derives from ancient Greek mythology and the Titaness "Theia," who gave birth to the goddess of the moon, "Selene," which, in turn, gave her name to the element selenium (Se). Isn't it beautiful how science and religion get along so incredibly well all the time?

of the newly molten planet Earth and into a semi-stable orbit* around the ball of liquid rock. The material that was ejected from this collision eventually coalesced into what we now call the Moon, producing a brilliantly tidally locked, lower-density-than-Earth extra-large satellite for us to marvel at 4 billion+ years later. To humans throughout history, the Moon has always been a convenient and nice object for individuals to look at while pondering the important (or unimportant) questions of the day. But now we know not only that we have a moon, but its process of formation was a major cause of why life, and thus humans, exist on Earth at all.

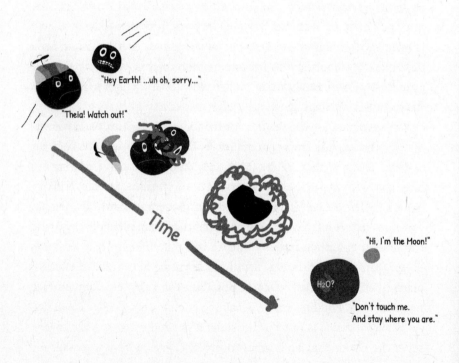

A nonexact play-by-play of how the Moon formed.

---

* Whereas some of the material from this violent collision would have been blasted off to eventually end up in the Sun, anything with a velocity lower than the escape velocity of the Earth would have still been gravimetrically tied to our planet and would have stuck around one way or another.

The Important Consequences of the
Moon-Forming Impact on Earth

Much like any one-way timeline that cannot be rewound and run again under different conditions, it is difficult to know what would have happened if X or Y had not occurred over Solar System history. We ended up where we are because of an innumerable string of events, some large, some small, but all contributing to the overall evolution and eventual grand habitat of planet Earth. What can be said with absolute certainty is that if there had not been a Moon-forming impact at the time it happened in Earth's history, things would have evolved very differently on our planet. Of course, it is *possible* that without this early jolt of excitement and high-temperature sterilization, intelligent beings would have developed on Earth a billion years earlier and Theia just hamstrung the development of hover-boards and space travel. But far more likely is that without Theia hitting the reset button on Earth ~150 million years in, things would have gone far less swimmingly for life to take hold at all, forever depriving the Solar System of things like Molly McButter and *Ace Ventura: Pet Detective*.

The instantaneous ecological consequence of Theia and Earth meeting was the immense heat produced, leading to an ocean of liquid-hot magma.* Whereas there is no evidence that life as we know it had started to grab a foothold on Earth prior to the Moon-forming impact, if it had, any progress made in that direction was almost certainly wiped out. Of course, it is *theoretically* possible that life could exist happily doing the backstroke in molten rock at 1500°C, but this life would have had to be in a completely different form from anything we have ever encountered or even consider remotely possible. The reason is, at such high temperatures, building and maintaining complex molecules like amino acids and proteins is just not possible. It is simply just too hot to preserve the bonds that hold them together. Consequently, the trailhead for any path that led to life as we know it must have started following the magma ocean stage of Earth's history.

---

* Originally coined by Dr. Evil, this term is now commonly used in geologic circles. At least I wish it was.

Not only did Theia's arrival mark a "resetting" from a biological and Earth surface perspective, it had a massive effect on the atmospheric evolution of Earth. The atmosphere of a planet and how it develops is a complicated balance between gases, gravity, temperature, and luck. This balance is perfectly revealed using our sister planet, Venus, as a comparative example. Venus is the second planet from the Sun: it is about the same size as Earth, it is made of the same basic material as Earth, and it is roughly the same distance to the Sun as Earth. Yet Venus has a crushing atmosphere that exerts pressure almost 100 times greater than Earth and hosts the hottest surface temperatures of any planet in the Solar System. If Venus ever had significant amounts of water, its runaway greenhouse atmosphere boiled all of it away on its journey to an inhospitable 96 percent CO_2 atmosphere that contains way more sulfuric acid than even a Batman villain would find comfortable. So if Earth and Venus are so similar in so many basic planetary ways, then why are they so vastly different in their atmospheres? The answer is likely because of Theia.

It is difficult to know exactly what the pre- or post-Theia impact atmosphere was like on Earth and understanding the evolution of Earth's atmosphere is a very active area of research. Many opinions exist about which gases would have dominated the early Earth, and many of the arguments stem from the various possibilities of what the Earth was originally built from.[*] Regardless of Earth's exact original building blocks, many scientists speculate that prior to the Moon-forming impact, Earth was on a similar atmospheric trajectory as Venus: a runaway greenhouse that would have culminated in wholly hostile conditions for development of anything other than a torture outpost for sadistic aliens. Irrespective of what our exact starting position was, the game was changed in a big way when Theia came to

---

[*] Building Earth from different types of meteorites leads to disparate outcomes for early atmosphere models. This largely has to do with the reduced or oxidized nature of the building materials, or the amount of available oxygen and other gases in the system.

town. Such an impact essentially reset the system—not only by blow-ing off large amounts of the atmosphere that had developed in the first ~150 million years, but critically, such a large impact would have liberated primordial gases from the Earth's mantle. Now here, it may seem reasonable to think that any volatile species, say, water, that was present on Earth either as an ancient ocean or water locked up in the early mantle would have been completely lost out of the sys-tem due to a massive superheat-producing impact. However, again this is space, and size *really* matters. Gravity doesn't care if a planet is liquid rock or solid gold, gravity permits the object to hold on to everything it can, and the significant gravitational pull of even a beaten-up Earth is enough to keep most of the volatile species close enough to where they will eventually settle back down onto the planet. As such, the result was a post-impact atmosphere rich in things like hydrogen, carbon monoxide, and water.* These chemically reduced ingredients are important because abundant chemical energy exists in a reduced atmosphere. A reduced atmosphere means there is ample chemical en-ergy available for any intrepid balls of organic goo that decide to spring to life/colonize Earth. The difference between a reduced atmosphere (hydrogen, methane, ammonia) and an oxidized one (carbon dioxide, inert gases) is like the difference in trying to develop and proliferate inside a virtually endless all-you-can-eat buffet of energy and trying to develop and proliferate on a barren slab of inhospitable rock.

So, with the conditions favorably reset roughly ~4.4 billion years ago, an opportunity (and available energy) emerged to turn raw, abiotic, carbon-containing molecules into what we term "life," however one chooses to define it. Perhaps all thanks to a chance encounter with an immense meteorite named Theia.

---

* The exact source of most of Earth's water remains unknown. It could have come from (1) the mantle of Earth that was liberated during the Moon-forming impact, (2) delivered from Theia, or (3) delivered by later arriving meteorites or comets rich in water. Or a combination platter of all of the above. We don't yet know.

Speculation About the Material
Delivered by Early Meteorites

The Moon-forming impact turned Earth into a giant ball of molten rock. This concept may seem like the natural outcome of such a lively collision, but it is important for a number of reasons. First, from a chemical perspective, when bodies are molten, geochemistry happens at a high rate. Elements have chemical preferences, and when everything is a liquid, it is far easier for elements to move around and satisfy their chemical desires. Just as people tend to segregate into similar cultures in big cities, approximately the same is true for the chemical elements in a planetary body. Very dense materials like iron and nickel metal like to move to the center of the body and form a planetary core. There are elements that desperately want to be near iron (appropriately called "siderophile" or "iron-loving") and when the Earth was molten, the vast majority of iron's element buddies like iridium and gold moved to the core of the planet, leaving only a vanishingly small percentage left in the crust.* Conversely, there are elements like magnesium, calcium, and aluminum that have no interest at all in living in the metal core and want to live out on the periphery of the planet (elements termed "lithophile" or "rock-loving"). These chemical suburbanites therefore make up the crust and mantle of the Earth. Because of this predictable chemical behavior, if, after the Moon-forming impact, things proceeded with no other addition of meteoritic material, then we would have essentially no access to any gold, iridium, or any other of the chemically similar elements because they would all be stashed in the core of our planet. But we do have some gold. We do have some iridium. And we have a whole host of other things in the crust that should all be in the core. Then why do we have them in the crust? It is because arrival of meteorites did not stop with the Moon-forming impact. And importantly, these raw materials arrived late to the party—after the Earth was done forming a core. As a consequence of the mantle and crust no longer being liquid

---

* For instance, >99.9 percent of Earth's budget of the element iridium following the Moon-forming impact migrated to the core.

rock, these poor iron-lovers became the chemical equivalents of Tom Hanks in the movie *The Terminal*. They are stranded in purgatory, being shuttled around the crust and mantle by plate tectonics, but unable to relocate to the core, where they want to live.*

There is little debate that humans have late-arriving meteorites to thank for our accessible budget of certain precious metals like gold, platinum, and iridium. And you can have a reasonable debate about how much society *actually* needs these elements, but since multiple empires have been built and destroyed in the quest for more of them, these shiny sojourners have had an undeniable influence on humanity. However, these fashionably late meteorites may have contributed much more than just precious metals to our crust: they may have delivered the organic† material from which life developed, as well as the water on Earth that sustains it. Not bad for a few less-than-punctual stones.

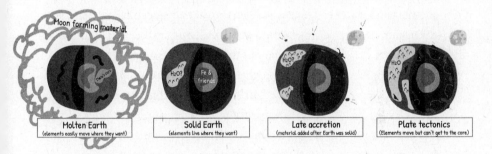

Arrival of material to Earth after the Moon-forming impact replenished some metals in the mantle and crust, and may have provided (1) lots of water and/or (2) basic organic materials.

---

* This has been termed the "late-accretion" or sometimes the "late-veneer." You may feel bad for the siderophile elements that can't make it to their preferred home, but the existence of these precious metals has certainly been beneficial to the jewelry industry, which is of course the backbone of any healthy society.

† The term "organic" is different from the term "biological," although the terms are often interchanged. "Biological" means it is related to a living organism, whereas "organic" just means a carbon-containing chemical substance. All biochemicals are organic, but not all organic chemicals are biological.

Continued Help from the Moon

Regardless of whether the ingredients for life, or life itself, was delivered by the late-accreting material to Earth, there is little dispute that organisms have populated the Earth for at least 3.8 billion years, as evidenced by microfossils in a number of different rocks. Less direct (and therefore much more disputed) lines of evidence hint that life on Earth may date as far back as around ~4.3 billion years—in other words, just after the Moon formed ~4.4 billion years ago. Beyond the immediate environmental construction assistance from Theia, the actual post-impact, fully formed Moon itself has been the subject of many thought experiments regarding the emergence of life on Earth. Many researchers would argue that without the Moon, Earth would have remained a boring lifeless rock, or at the very least, life would have taken a very different path than it eventually did.

Quite possibly the most important consequence of the Moon on the history of early life is the same important consequence it has on modern maritime life on Earth: the tides. Modern tides range from ~zero to more than 16 meters, which, depending on your exact location along the coast, may seem like a lot of water-height variability. Modern high and low tide both occur ~twice each in twenty-four hours. However, 4 billion years ago, Earth was spinning much faster and the Moon was much closer to us. These differences caused much larger tide fluctuations to occur more frequently: there might have been ~50-meter tide changes every five hours or so in certain places. You may be wondering what this has to do with the development of life? Well, according to many origin-of-life researchers and astrobiologists, tides could have had a lot to do with early life on Earth. In order to create that lovely broth of interacting organic molecules lovingly termed the "primordial soup," it is important to concentrate organic molecules to the point where they interact, or in other words, to get rid of a lot of water. One easy way to get rid of a lot of water is to evaporate it, but in order to get that primordial soup recipe just right, you would likely have to try over and over and over again. The most efficient way to repeatedly do this is not an exotic process, it is just a natural consequence of oceanic tides. Ocean tides produce local differences in the water level on a repeated basis, which happens to be the perfect mechanism for concentrating organic material. When the tide is high, water that contains dilute random organic molecules is repeatedly tossed up onto rocks, often forming small tidepools. When the tide recedes, the trapped water evaporates, concentrating the organic material that was in the water. Because this sequence repeated multiple times a day, day after day, all around the world, tidal areas would have represented a great place for prebiotic organic material to randomly interact, making the jump from lifeless goo to biotic goo. Without the Moon and the tides it creates, this crucible for carbon concentration would barely exist.*

A thorough discussion of the Moon and its importance on the development and evolution of life is far beyond the scope of this book. There are numerous reasons why the Moon has played a crucial role in the development of complex life. For a comprehensive expansion of this topic, check out work by people like Dr. Richard Lathe. However, if you are looking for a less academic and more salacious read about the lunar influence on modern *human* life, then familiarize yourself with the story of Thad Roberts, a disgraced intern at NASA who conspired with fellow interns to steal lunar samples collected by Apollo astronauts. In short, following their successful (and actually quite impressive) heist of ~8 kg of Moon rocks, Thad and his love interest Tiffany decided to spread the lunar material out on a bed and have sex on top of it† before trying to sell the postcoital stones to interested buyers. Unfortunately, at least for the lunar sex squad, the potential procurers of the material were members of the FBI running a sting operation to recover the precious stolen material. It is safe to say that following the . . . uh . . . excitement, those particular samples became slightly less valuable from a research perspective. But an interesting story, nonetheless.

---

* I want to be clear, while scientists don't know that life sprang up for the first time in a tidal area, the oft-repeated natural concentration of carbon compounds that occurs in tidal zones makes them one of the most likely areas for it to have happened.

† Hence, the title of the critically destroyed book *Sex on the Moon,* which documents, apparently poorly, the theft and attempted sale of the otherworldly aphrodisiacs.

More Recent Large Meteorites and
Their Influence on Earth History

Even though plate tectonics—Earth's continent-size recycling program—has done an impressive job of erasing and replacing large portions of the rock record over geologic time, geologists are still able to
find scars of meteorite impacts as far back as 2.4 billon years.* The second-
oldest known crater, the Vredefort crater in South Africa, is just over
2 billion years old, and with a crater diameter of ~300 kilometers, is the
largest surviving known impact structure on Earth. However, almost
certainly, older and likely larger craters were present at some point on
Earth's surface but have just been destroyed or overprinted by tectonic
activity and/or weathering. It is also important to consider that Earth's
surface is about 70 percent water, and as such, only 30 percent of its
historical impacts have even a semi-decent chance of being preserved
at a reasonable level. The significance of impacts on Earth, identified or
not, large or not, is multi-fold. For one, they show that we are routinely
pelted by flying rocks of a variety of sizes and have been for billions
of years. Second, this constant bombardment is not just an interesting
side note about Solar System processes and gravitational dynamics, but
it is also of paramount importance to the plants and animals that are,
and want to remain, alive. As it turns out, when a large rock hits Earth
at say, 40,000 miles per hour, or ~25 times faster than an average bullet,
it can make quite a mess of things.

Perhaps the best-preserved and most easily accessed example of the
type of damage that can be done by something, anything, moving at
such incredibly high speeds is found at Meteor Crater in northern Arizona. The crater has a diameter of about 1.2 kilometers and is thought
to have been caused by an impactor about 50 meters in diameter, or
about the size of the visitors' center that is barely discernible in the
accompanying photo (the black thing is the much larger parking lot,
so that doesn't count). It is impossible to stand at the rim of the crater

---

* This is the ~16 kilometers diameter Suavjärvi crater in Karelia, Russia.

Top: *An overhead view of meteor crater and its visitor center in northern Arizona.*
Bottom: *Intrepid geologists investigating the crater up close.*

and not think about the sight when the impact happened ~50,000 years ago. When one realizes that Meteor Crater is not considered a very large crater on Earth, this becomes just a tickle scary.*

Luckily for us and all other living things, the frequency of the impact is inversely proportional to the size of the event. This is a fancy way of saying that the small ones are common and the big ones are rare. Earth takes quite a few tiny pokes and the occasional jab, but only very rarely does Earth take a planetary haymaker on the chin. The tiny pokes and jabs can of course have devastating effects on any local life in the neighborhood, but little effect on more far-flung colonies of critters. Conversely, the rare very large events can have a devastating effect on the entire planet when they happen—think superheated atmosphere, multi-kilometer high mega-tsunamis, and global atmospheric dust blotting out the Sun for years at a time. Lucky for everything that has ever lived on Earth, very large meteorite impacts, ones that produce craters >100 kilometers in diameter, are exceedingly rare events. Only five such events are known in the last 2 billion years, and only three of those events occurred after the point when Earth's creatures were evolved enough to have backbones.†

Paleontologists spend a lot of time studying the rock record and documenting the history of life on our planet. What comes up time and time again is the propensity for life to chug along fine, slowly and happily evolving into more complex organisms, until some environmental disaster happens and, presto, mass extinction. Since this is a book about meteorites, you may see this as the point where all of the major mass extinction events in Earth's history are tied to meteorite impacts. Well,

---

* To provide a bit of scale, the energy released when the Canyon Diablo meteorite struck this site was approximately 650 times more than the atomic bomb dropped on Hiroshima, Japan, at the end of World War II. So, yeah, meteorite impacts release a hell of a lot of energy.

† Backbone evolution occurred during what is termed the "Cambrian Explosion" just over 540 million years ago when life on Earth was in its "college years." Creatures were experimenting, rapidly evolving, and figuring out new things like how to best build a body and where to get the best weed.

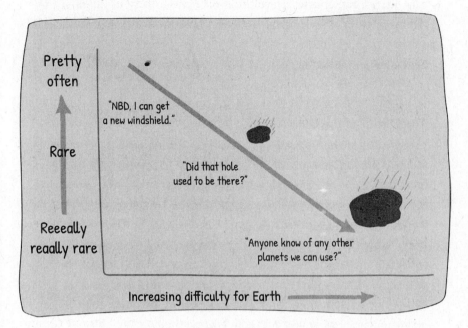

that would certainly be a sexy segue into how space rocks are constantly destroying and reshaping our biosphere, but it would not necessarily be true. Of the "Big Five" mass extinctions endured on Earth, actually only one enjoys a consensus view as being caused by a meteorite impact.* There are a couple of minor mass-extinction events that have been linked to meteoritic impacts,† and these smaller events have undoubt-

---

* The direct causes of three of the other four are unknown, actively debated, or appear to be a combination of multiple environmental stresses. Alarmingly, things we are currently experiencing such as rapid sea level and global temperature changes are heavily involved in almost all mass-extinction events in geologic history. Gulp.

† One such "minor" event is the Eocene-Oligocene extinction (~34 million years ago), which has been tied to the Popigai crater in Siberia. This event eradicated >10% of all marine life in a blink of geologic time, making a term like "minor mass-extinction event" difficult to digest given the sheer amount of death that occurs. However, the term "minor" is used because the Eocene-Oligocene extinction is not even one of the more widespread and deadly mass-extinction events in Earth's history. Double gulp.

edly altered the course of evolution over the history of life on Earth. However, there is only a single meteorite impact principally linked to a major mass extinction event, and you probably already know which extinction event that is.

The Death of the Dinosaurs

If you were growing up in the 1950s and '60s, the Moon, Mars, or anything at all to do with space was undoubtedly the topic that captured your imagination and the imagination of the greatest amount of kids around the developed world. Of course, the two world superpowers then, the United States and the Soviet Union, were engaged in the famed "Space Race," which consumed almost all the oxygen from most newscasts and conversations at the time. Although this race was more about politics, pride, and military prowess than academic curiosity, it still inspired the hell out of a lot of young girls and boys, men and women alike, to look up and envisage what was out there beyond the horizon.

However, if you were growing up in the 1980s and '90s, space had a serious contender for your and countless other kids' attention in the form of dinosaurs. Kids' movies like *Baby: Secret of the Lost Legend* and *The Land Before Time* were evidence of Hollywood attempting to cash in (and helping feed) dino-hysteria. The pre-history fever pitch was evident at the beginning of the 1990s with the understandably intense popularity of Michael Crichton's *Jurassic Park* and the less understandable popularity of ABC's sitcom about a blue-collar family of mildly sentient dinosaurs, ingeniously titled *Dinosaurs*. What was driving this public craze for stories and information about the legendary lizards that used to rule the Earth? I would like to think it was science; there is really only circumstantial evidence for that assertion, but hear me out. Around 1980, when Mount Saint Helens was explosively reminding people of the importance of geology and the power of Mother Nature, scientists were making extremely important discoveries and developing theories that would captivate the public like only a few things can. Terms like "geology" and "excitement" are in most times considered by most humans to be oxymoronic, but when you roll together geologic spectacularities like volcanic explosions, giant meteorites hitting Earth, and extinction of the dinosaurs, people pay attention.

Prior to around 1980, it was unknown what caused the extinction of the dinosaurs. At the time, this had to have seemed like a fairly large knowledge gap, and we should probably figure out what vanquished the hundreds of building-size lizard species that roamed unchecked around the planet. Why? Well, knowing about their extinction could prevent us from being vanquished ourselves by whatever it was that took out the dinos. And if they ever came back, it sure would be nice to know how to defeat them. Regardless of the exact motivation of any individual, it seems it should have been, and was, important for a number of reasons to a large number of people. Theories of what caused the dino-demise were plentiful, and many of the less scientifically rigorous ones bordered on absurd. These ranged from increased global temperatures causing all of the dino-offspring to be males, very hungry caterpillars eating all the vegetation, to an especially virulent strain of dino-herpes decimating the population to the point of extinction.* However, around 1980 the first hard evidence started to emerge about what likely caused the mass extinction of the planet's previously dominant creatures. The father-son team of Luis and Walter Alvarez discovered a thin layer of sediment that was spread around the planet and had an especially high concentration of the element iridium. This may not sound like a big deal, but as discussed above, the element iridium is found in extremely low concentrations in the crustal rocks of Earth yet is found in very high abundance in, you guessed it, meteorites. It just so happens that this iridium-rich layer also separated rock layers that contained very different types of fossils, suggestive of a very abrupt change in the biosphere.

After this, the story just writes itself, right? Well, scientists are an excitable but careful bunch. An asteroid hitting Earth and killing the dinosaurs was a scientifically sexy idea, but it took considerably more

---

* I made this last one up, but if you are interested in some of the actual proposed reasons, weird or not, about how the dinos died, they are chronicled, along with many other interesting things, in Brian Switek's 2013 book, *My Beloved Brontosaurus*.

The father-son team of Luis (at left) and Walter Alvarez (at right) posing at the iridium-rich layer near Gubbio, Italy (tilted subsequently by tectonics). Walter's right hand is touching the last layer of rock during the dinosaur's time. Luis's left hand is touching a dino-free existence.

evidence than a thin, exotic metal-rich layer of sediment in the right position in a few places around the world to arrive at the consensus view that we have today. After all, geology was the study of slow change over long periods of time; religious texts like the Bible were where sudden catastrophic events belonged, right? As more studies followed looking at rocks the same age as those investigated by Family Alvarez, more and more locations were exhibiting the same results. In the rock layers below the iridium-rich layer, there were dino bones. In rocks deposited after the iridium-rich layer, the dino bones were absent. And because the data were similar regardless of where on Earth you looked, this pointed toward a global phenomenon, not just a localized event. When the iridium-rich layer was examined in microscopic detail, it contained further, strong evidence pointing toward a meteoric impact ~66 million years ago. First, the layer contained impact-formed microspherules, tiny rock spheres that form when giant impacts happen because some of the rock is molten and tossed into the air; second, it also contained impact-shocked minerals that recorded imprints of the impact in their partially damaged structures, like how you know someone T-boned your Honda Civic because the side is caved in. As work continued on what is one of the most important rock layers in Earth's history, evidence such as thickness of the layers and locations of where impact-shocked minerals

The death of the mighty dinosaurs and the rise of the furry little mammals.

were found increasingly pointed toward Central America as the epicenter of the impact. The eventual discovery in 1990 of a large impact crater in Mexico with the same age as the iridium-rich sediment layer was the final bit of evidence that most scientists needed: a meteorite killed off the dinosaurs,* forever altering the history of our planet.

---

* Even though this is a book primarily about meteorites, it is incumbent upon me to mention that there are vocal (and very reasonable) detractors from this consensus view of the cause of the end-Cretaceous extinction. The eruption of the Deccan Traps, a very sizable outpouring of rock that now covers a vast area of what is now India, dates to the same time as (and possibly directly at) the end-Cretaceous extinction. This massive eruption would have provided deadly amounts of noxious gases to the atmosphere and certainly would have contributed to the extreme difficulties that the dinosaurs faced at the end of their reign. Some researchers say the eruption could have been the primary

The impact—in both meanings of the word—of the meteorite that caused the death of greater than 75 percent of the species on Earth is difficult to comprehend. The approximately 10–15 kilometers (6–10 miles) wide asteroid was moving at a speed of around 25,000 mph when it landed in a shallow Caribbean Sea overlaying carbonate and sulfate rocks. The modern physical scar is the 180 kilometers diameter Chicxulub impact crater in the Yucatan region of Mexico, but the effects were far from only local in scope. The immediate repercussions of such a large impact have been modeled extensively, and while there is considerable debate about the specific aftereffects, scientists agree that it would have been very bad news for life that was there at the time. Mega-tsunamis would have inundated anything anywhere near a coastline. Many models suggest that the atmosphere was superheated to the point of igniting raging forest fires around the planet, only to be followed by severe global cooling due to particulates blocking out the Sun for months after the event. Because of where the impact happened, fun things like global acid rain would have been commonplace thanks to the impact vaporizing the sulfate deposit on which the asteroid landed. Earth was simply not a pleasant place to be in the aftermath of being smacked by a giant rock moving incomprehensibly fast. Thankfully, however, all was not lost. As Jeff Goldblum is famous for saying in *Jurassic Park* (and the million memes that followed), "Life, uh, finds a way."*

---

cause of the dino-demise. Other researchers believe that the combo of a supersonic rock from above and outpourings of rock and gases from below were the dueling banjos of death for our dino friends. After all, perhaps the primary reason mass extinctions are so rare on Earth is that it takes multiple global stresses at once to outpace life's ability to evolve and adapt. Geologists continue to unravel the order and magnitude of the known stresses and may find other stresses in the future. But virtually every researcher agrees that a huge speeding rock hitting Earth was probably not a positive thing for life in the dinosaur community.

* Perhaps better quotes would be from Sun Tzu in *The Art of War*, "In the midst of chaos, there is also opportunity," or from Petyr Baelish in *Game of Thrones*, "Chaos is a ladder," but the harmony of using *Jurassic Park* was too much of a temptation.

The Upshots of a Decimated Planet

By any metric, dinosaurs were a very successful branch of animals. They developed, evolved, and diversified over the better part of 175 million years. Some were gigantic, up to 50 meters long, and some were barely the size of a chicken, but together they were the keystone organisms in the food web for the bulk of the Mesozoic Era. Whereas evidence exists that dino domination of the planet was starting to wind down even before the death blow at Chicxulub, there is little question that things changed in a very big way following the Cretaceous–Paleogene (K–Pg) extinction event ~66 million years ago. This is especially evident in the fossil record when you consider that nothing with a body weight over ~25 kg survived out of the Cretaceous period. Think of that: one moment there were creatures as big as 100 tons strutting their stuff around, and then, in a blink of geologic time, no living animal on Earth was larger than a basset hound. The dinosaur departure, irrespective of an instantaneous or slightly protracted one, created immense opportunity for previously stifled species. And who better to capitalize than those furry little opportunists, the mammals.

Mammals didn't just spring up after the lizard kings were dethroned. The first mammals evolved alongside their reptile overlords more than 200 million years ago. But as you can imagine growing up in the long shadow of the giant lizards, those Mesozoic mammals evolved primarily as small nocturnal creatures scurrying about trying not to be noticed as the perfect warm and fuzzy snack. However, the mammalian species that managed to survive the K–Pg extinction were the beneficiaries of a completely new world of possibilities. Well, it was a positive at least that there was a lack of giant lizards around every corner trying to eat you, but there were still a lot of food shortages due to large chunks of the planet being burnt and fallow, but the "no giant lizards" part was surely a relief. With this newfound freedom, mammals flourished, to say the least. At the start of the Paleogene period, the mammalian class underwent a massive increase in numbers and body size from scroungers the size of gerbils to the modern blue whale, which is the largest animal ever known to exist. Mammals diversified exponentially, and it was not

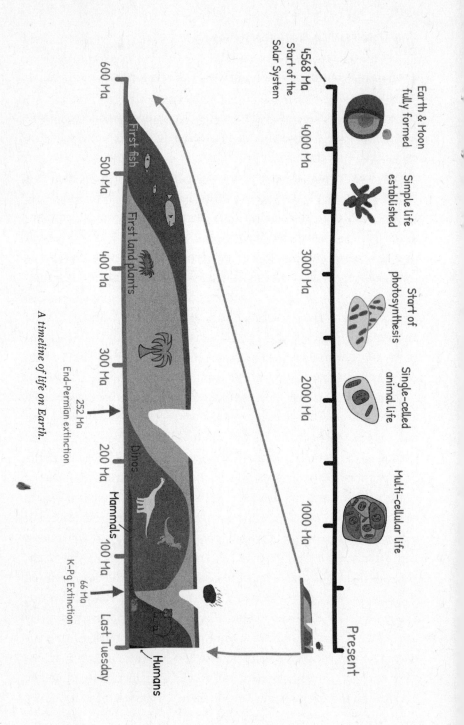

4568 Ma
Start of the
Solar System

Earth & Moon
fully formed

Simple life
established

Start of
photosynthesis

Single-celled
animal life

Multi-cellular life

Present

600 Ma

500 Ma

First fish

400 Ma

First land plants

300 Ma

252 Ma
End-Permian extinction

200 Ma

Dinos

Mammals 100 Ma

66 Ma
K-Pg Extinction

Last Tuesday

Humans

4000 Ma

3000 Ma

2000 Ma

1000 Ma

A timeline of life on Earth.

long until they were flourishing not only on land but also in the sea and in the air. We as *Homo sapiens* arrived on the evolutionary timeline after mammals had been around for >200 million years. However, *by far* the most productive of those evolutionary years were the last ~66 million, where mammals experimented and evolved freely without having to worry about staying hidden all the time from their terrible lizard oppressors.

Regardless of whether it is viewed holistically or as one occasion at a time—events like the Moon-forming impact, the late-accretion of raw and unprocessed materials to early Earth, the periodic pounding of smaller meteorites, or the extinction-causing impacts shifting evolutionary history—it is not hyperbole: humans owe their very existence to meteorites, and we are just getting started on the reasons.

Two

Cosmic Cinema for Early Humans

Astronomical displays, such as eclipses, cometary flybys, or distant supernovae, have been happening intermittently in the sky since our planet formed. These events have occurred throughout the entirety of Earth's history, and there have undoubtedly been some really impressive shows of cosmic power over the last 4.5 billion years—the dinosaurs might have one hell of a story to tell if they were more communicative, for instance. Unfortunately, however, historical accounts of such celestial cinema is limited to the big-brained, yet very peculiar humans that came to be only in the latest flash of geologic time. Many of the earliest human/cosmos interactions are lost to history (due to a lack of records, or even further back, a lack of written/spoken language). But even without the benefit of records spanning all human history, we must have been noticing things happening up in the sky for a long time. Looking up eventually became a tool to aid early human navigation, but early on, the sky was probably just thought of as one big shiny light during the day and numerous small, twinkly lights to look at when nighttime set in. Day to day and night to night, the sky may seem like a static thing, but to the careful observer, subtle things happen beyond just the movements of the Sun and Moon. Every so often, though, something very dramatic occurs in the sky and even the most oblivious primates take note. These dramatic events, whether a cometary arrival, a faraway exploding star, or a nearby meteoritic airburst, have helped shape our culture to what it is today.

For most of human history, the bulk of the human population has been concerned exclusively with day-to-day activities and the happenings on the ground. And this is completely understandable. But things do happen off the ground every so often, and sometimes they cause quite a stir. Just try for a bit to empathize with a cabbage farmer whistling and working the fields three thousand years ago when, suddenly, the Sun completely disappears in the middle of a cloudless day. You will likely have no trouble seeing how such an unexpected event would cause a bit of unrest in the mind, and if Mr. Cabbagesmith was the excitable type, you could see how that mind might venture into territory that could be described as "hysteria."

Until recently in human history, there was not an overabundance of appreciation for what was happening in and above the sky from a scientific perspective. This is probably why, when something like a solar eclipse happened, people sometimes lost their cool. Now, of course, thanks to centuries of study and TV documentaries hosted by Neil deGrasse Tyson, we realize that a solar eclipse occurs when the Moon gets between your little spot on Earth and the Sun (with a rearranged order for a lunar eclipse). Eclipses are predictable—to the minute—decades in advance and yet they still garner bounteous news coverage and eyewitness accounts whenever they occur. Likewise, we basically know when and what to expect for luminous cometary flybys, and yet they still create a massive worldwide buzz when they happen. These things are proof of how much influence cosmic events can have on humans, even in the modern age when we think we know what is happening and why.

A noteworthy example of the, er, curious human response to such an event can be found in the first known recorded reference to an eclipse. This comes from Loughcrew, Ireland, in the form of three stone monuments that host carvings of spiral petroglyphs. Based on back calculations of previous eclipses that would have been visible in the area, the depicted-in-stone alignments of the Sun, Moon, and horizon likely represent the solar eclipse that occurred November 30, 3340 B.C.E. However, what remains unclear from the carvings, and perhaps understandably omitted by the ancient artist/authors, is why there is a mass grave hosting the burned bones of four dozen people in the area. Some might suppose it has something to do with human sacrifice to a god that somehow was connected to the Sun. Or, it could just be one of those crazy coincidences of history where scores of charred bodies turn up disturbingly close to an ancient observatory for no apparent reason.

The happenings at Loughcrew are not a singular story of humans . . . overdramatizing . . . an eclipse. Intense reactions and creative setup scenarios seem to be the norm and not the exception. In many ancient (and present) cultures, eclipses are representations of battles between great

Solar eclipse. Time to freak out.

or divine beings, with various iterations depending on the culture. For instance, Vietnamese legend has it that a giant frog is devouring the Sun. The Vikings saw eclipses as a deadly game of chase involving two sky wolves trying to eat the Sun and plunge the Earth into eternal darkness.* The ancient Chinese thought the Moon was a dragon attacking the emperor (projected as the Sun). And since any deft emperor would want to avoid being attacked by a dragon, emperors generally employed several astrologers to help predict such events. Famously, in 2300 B.C.E., two of the emperor's finest failed to get the prediction right, resulting in their beheadings. Which is a bit unfair, since—and this is just speculation here—their failed prediction did not result in the emperor being devoured by a giant sky-dragon.

And speaking of headless, Hindu belief interprets an eclipse as the detached head of the deity Rahu occasionally ascending into the sky after it was removed from his body. Importantly, while Rahu's head achieved immortality, the rest of his body was not so fortunate, as the two were intentionally separated by his foe the very moment he was drinking an immortalizing nectar. Consequently, the now-eternal head of Rahu is trying to seek revenge on the snitches (the Sun and Moon) that led to his beheading. As the lore goes, Rahu is able to eat the Sun (or the Moon in the case of lunar eclipses), but since he has no body, they just pass through his throat and back out the other side unscathed. As the old saying goes, "Revenge is a dish best served up at recurring intervals that happen to coincide with the predictable motions of the celestial bodies."

The ancient Babylonians were able to vaguely predict eclipses, but nonetheless saw them as bad omens, particularly for the health of their royalty. As such, when an eclipse was predicted, a low-life commoner would be given the "honor" to sit atop the throne just in case something bad occurred that was aimed at its usual occupant. After the eclipse was over, the stand-in was rewarded for his service by, you guessed it, promptly being executed.

---

* If you have ever spent any time in the northern European stomping grounds of the Vikings in the wintertime, eternal darkness is a very understandable and legitimate thing to fear.

The ever-uplifting ancient Greeks saw any eclipse as a certainty of doom—although things did not always turn out so gloomy when eclipses actually happened. In perhaps one of the stranger occurrences in military history, the unnatural twilight of 585 B.C.E. reportedly stopped the sparring Lydians and Medes cold in the middle of an ongoing battle in what has gone down in history as the "Battle of the Eclipse." Allegedly, both sides interpreted the eclipse as an omen to lay down arms, ending more than a decade of warring and leading to a peace treaty. Since that treaty was signed, there has not been, nor will there ever be, any more political tension in the Middle East . . . all thanks to that eclipse.

As you may have guessed, military responses to other solar eclipses have not always been so pacifistic. In what is arguably the most famous historical reaction to a solar eclipse was the northern European totality, or full eclipse, that coincided with the death of King Henry I of England in 1135. Apparently, this eclipse also darkened the hearts of men, as the struggle for the newly vacated throne threw the kingdom into almost two decades of chaos and civil war.* If history is any guide, this turmoil may very well have happened without the disappearance of the Sun, but it is of course more fun to blame things on astronomical events and not just the commonly detestable nature of human beings and the struggle for dominion. As a fun historical note—and likely not a big-time shock to a thoughtful reader—it was actually not the eclipse that killed the king. Henry insisted on eating a number of lamprey during a foray in northern France, where such creatures were a local delicacy. Oh, Henry . . . after his meal, he fell terribly ill over the following week and eventually succumbed to the icy grip of death. Tough break when your last meal is a pile of slimy jawless fish parasites, but death comes in many flavors.

When reliable eclipse predictions first became possible, only a few learned or fortunate people were privy to the information. In fact, America's distasteful "discoverer" Christopher Columbus put early astronomical tables to use in a story that very much summed up his char-

---

* This period of chaos is affectionately termed "the Anarchy" by modern-day historians.

Henry I of England, the king dethroned by jawless fish parasites.

acter. In the summer of 1503, Columbus haplessly stranded himself and his merry men on the friendly island of Jamaica. For about six months, the kind locals were extremely accommodating, providing food and provisions for the stranded seafarers. However, by around March 1504, the local leaders decided that Columbus and the castaways had started to overstay their welcome and chose to stop supplying the freeloading vagabonds. Columbus found a solution to his provisions problem in an almanac aboard his ship. The almanac contained astronomical tables covering the years 1475–1506, and just so happened to provide him with the date of a lunar eclipse in the area. On the eve of the predicted eclipse, Columbus called a meeting with the tribesmen that had been previously so gracious, stating to the leaders that their treatment (or current lack thereof) of he and his men was angering his god and they would receive a clear sign of this divine fury that night. On schedule, the lunar eclipse occurred, so impressing and frightening the Indigenous Jamaicans that they promptly ran with ample provisions as to not incur any further wrath of the apparently angered god. Columbus, the good guy he was, happily went on to subjugate his former hosts.

Eclipses can also be used in non-nefarious ways, and those with a scientific bent have studied them since ancient times with important consequences. Using eclipses, Chaldean (neo-Babylonian) astronomers figured out what is now called the Saros cycle—the approximately 18-year, 11-day, 8-hour interval in which the Sun, Earth, and Moon habitually line up for an eclipse. Due to that troublesome "8-hour" bit, the position on Earth where the eclipse is fully visible changes, meaning different parts of our planet get to freak out at different levels at different times, which is nice for parity's sake. It was lunar eclipses that allowed the ancient Greeks to accurately determine the distance between the Earth and the Moon, and the same events convinced Aristotle that Earth was a sphere due to the circular shape of its shadow when seen on the Moon. Please note that hard evidence that Earth is a sphere existed more than three hundred years before the time of Jesus. This is a very simple and clear observation that can be made by anyone, yet this is not enough to convince members of the Flat Earth Society they are clinging to ideas that have been obsolete for more than 2,300 years.*

Eclipses have continued to be a boon to science even in more modern times. And, no, sadly this does not reference the time when then President Donald J. Trump chose to look directly into the summertime North American totality of 2017 without any eye protection and decided to double the budget of the National Science Foundation.† Digression aside, a solar eclipse allowed the testing of Albert Einstein's then recently unveiled theory of general relativity. Einstein had suggested in his landmark work that massive objects caused distortions in space and time, and Sir Arthur Eddington confirmed this when he was able to

---

* In 2017 (A.D., just to be clear) basketball star Kyrie Irving publicly announced he believed the Earth was flat. Cleveland traded him shortly thereafter, but it is unclear if it was a direct result of his fleartherism.

† Only one of those things happened.

Donald Trump ~~doubling the NSF budget~~ *looking directly at the Sun during the 2017 solar eclipse.*

measure the ever-so-slight deflection of starlight by the Sun during the May 29, 1919, eclipse.

The above are just a smattering of the numerous eclipses that have happened over human existence, and only a handful of stories associated with them. An event like a solar eclipse is of course spectacular, but really all that happens is that it gets dark and maybe a bit colder for a few minutes in the middle of the day. A bit unsettling, yes, but it gets dark every night, so it just kind of happens at the wrong time. A little darkness is not something that is unfamiliar to almost anyone who has been on the planet for more than twenty-four hours. Now let's contrast a little untimely darkness with a giant fireball producing a light brighter than anything you have ever seen flying through the sky, possibly at you, accompanied by a thunderous noise louder than anything you have ever heard. On the "Should I Panic" scale, most would put "gigantic supersonic fireball" somewhere up pretty near the top, possibly even above seeing a circus clown with a solitary balloon walking slowly around the sewers of their hometown.

In any event, given how history has reacted to comparatively dull eclipses, it is understandable that the likes of supernova, comets, and especially large meteors have caused society a bit of consternation over the years.

Supernovae in History*

Supernovae, or exploding stars, are rare but massively energetic events that alter entire areas of the galaxy in the blink of an eye. Luckily for us, the galaxy is a massively large place, and when one of these rare events occurs, even at great distances from us, Earth can get a cool new sky decoration for an extended bit of time. In the year 185 C.E., Chinese astronomers recorded the arrival of a "guest star," marking the first unambiguous human observation of a supernova. The cosmic fireworks of the now far-less-cleverly-named "SN 185" were visible for an incredible eight months, and the remnants of the explosion are still studied by modern astronomers.

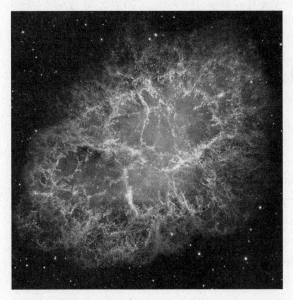

While no longer visible with the naked eye, the supernova recorded by Chinese astronomers in 1054 left a beautiful scar in the night sky in the form of the Crab Nebula.

---

* Just as a reminder, "supernova" is the singular and "supernovae" is the plural form of an exploding star. I here suggest that the associated collective noun should henceforth be a "blast" of supernovae. The English language is just so fun.

On multiple occasions since SN 185, supernovae have been noted by cultures around the globe but perhaps surprisingly, none of these celestial events have been directly tied to any wars or mass graves. However, particularly impressive supernovae in 1006 and 1054 C.E. have been credited with inspiring significant amounts of the rock art found in the American Southwest, and the event in 1572 was used by astronomical legend Tycho Brahe as important evidence that the heavens were not immutable, so there is that.

Comets in History

From the beginnings of recorded history, and likely before, people have been looking up and noticing things. And because comets—large balls of ice, rock, and dust that orbit the Sun with eccentric orbits—often produce a massive fiery tail visible even during the day for weeks at a time, they can be difficult to miss and were probably one of the first astronomical objects to garner attention. Many people have described comets in many different ways over the years: they have been known as "hairy" or "bearded stars," "demon stars," "broom stars," and a personal favorite, "menaces of the Universe." Regardless of whatever cute/terrifying name various cultures assigned them, comets were often considered harbingers of doom for those that were unlucky enough to gaze upon them. Perhaps it was because they disrupted the beautiful order of the sky, or perhaps it was just an easy way to get people to pay attention to you by spreading fear for no tangible reason. After all, ancient cultures did not experience extensive immigration, so humans had to use other divisive tactics to freak the populace out for personal gain.

Some archaeologists suggest that certain rock paintings and carvings, some dating to at least the second millennium B.C.E. and possibly before, portray comets. And whereas the images found in places like Toca do Cosmos in Brazil, Traprain Law in Scotland, Gegham Mountain in Armenia, as well as several other sites across the globe, certainly seem to portray what we know today as comets, without additional supporting evidence and written documentation, it is difficult to definitively say the ancient images are sketches of specific cometary encounters. However, comets definitively do play a role in the Babylonian *Epic of Gilgamesh*. In this ~4,000-year-old influential mythological work, the arrival of a comet brought the combo of fire, flood, and brimstone upon the people, so it is understandable why comets caused future consternation among the masses.

The first unambiguous renderings of comets come from writings and drawings on silk sheets from the second century B.C.E. China (Han dynasty), recovered from a tomb at Mawangdui. Recorded on the silk is a series of drawings of a variety of comets (or to the recorder, "demon stars"), accompanied by what each style of comet means. This impressive catalog is based on the appearance and character of the colors, size, and cometary tails. Spoiler alert: as you may have guessed, they are rarely good omens, regardless of what they looked like. Of the twenty-nine drawings and their reported "meanings," most of the translated implications of comets were things like:

General dies
War, famine

And not to dismiss the death of a general or famine for anyone, but some comets heralded even worse-sounding fates for the Chinese, such as:

Disease in the world
Calamity in the State

Some comets, while still interpreted as devastating, apparently arrived with a touch of generosity as well, as they were cataloged as responsible for odd combinations like:

Internal war. Bumper harvest.
Small war, corn plentiful.

And my favorite:

The small man cries.

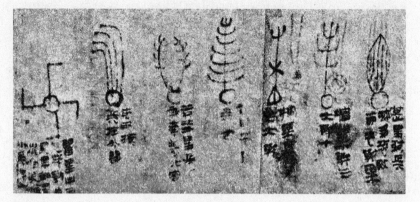

Comets on silk.

Comets have played an important role in Western culture and religion as well, perhaps most significantly with the appearance of one around the birth of Jesus Christ. It is widely speculated that a comet that appeared around 5 B.C.E. was the famed "Star of Bethlehem." This cometary arrival is referenced by multiple historical sources in the area and is also well documented by Chinese astronomers. And since a comet is clearly depicted in Giotto di Bondone's famous painting *Adoration of the Magi* (painted over 1,300 years after the fact), it is settled. Why would Giotto paint it that way if it didn't happen that way? Exactly.

Shortly following the death of Jesus, the Roman emperor Nero conveniently interpreted the appearance of a comet in 60 C.E. to justify the massacre of many of his political enemies. This preoccupation was one of his favorite pastimes when he was not busy burning sections of Rome to the ground and destroying historical artifacts, of course.

One of the best-known (and most influential) comets in history is Comet Halley. This comet was first definitively observed and recorded ~240 B.C.E. and visibly returns every 74 to 79 years. Every appearance of

Giotto's Adoration of the Magi *with a representation of a comet in the background.*

Halley has likely caused some interesting things to happen in the world, but some turn out to be more momentous than others. William the Conqueror took the impressive arrival of Comet Halley as a sign that he and his squad of Normans were going to put the smackdown on the English, resulting in one of the more consequential battles in the history of Europe. And as a test of consequentiality, go up to any student with a British accent and say nothing but "Battle of Hastings" and you will almost certainly, without hesitation, get the reply "1066."

In 1222, the return of Comet Halley and its reportedly spectacularly bright westward trajectory is said to have caused Genghis Khan to launch his invasion of southeastern Europe, leaving millions of people dead in his wake. However, it is likely the Khan-man was already planning to conquer the area, because, well, (1) it was area, and (2) it had women, two things that GK seemed to desire in great abundance, so this test of cometary consequentiality is dubious, at best.

Today, most people are familiar with the existence of comets at least, and even if they have not seen one themselves, they are likely aware of the amazing (and true) trivia nugget that American satirist and writer Mark Twain was born and died on occurrences of Comet Halley. For the uninitiated, Twain was born two weeks after the comet's closest approach to the Sun, or perihelion, in 1835. But of course Twain's cometary connection did not end there, and it was not lost on him in the least. In his autobiography, published in 1909, he said:

> I came in with Halley's comet in 1835. It is coming again next year, and I expect to go out with it. It will be the greatest disappointment of my life if I don't go out with Halley's comet. The Almighty has said, no doubt: "Now here are these two unaccountable freaks; they came in together, they must go out together."

Luckily? for Twain, he died the day following the perihelion of the comet in 1910, almost certainly with a smile on his face.

A less heartwarming modern tale of death and comets involves the comet Hale-Bopp and the Heaven's Gate cult of San Diego in 1997. Led by Marshall Applewhite, the members of Heaven's Gate were convinced that behind the luminous tail of Hale-Bopp was a spaceship that could carry them to, according to Applewhite, "a level of existence above human." To get on the spaceship, however, the members had to shed their Earth-bound bodies by ingesting a lethal mix of phenobarbital, applesauce, and vodka. The tragic result was a thirty-nine-member mass suicide. Applewhite additionally insisted they had to die with exactly $5.75 for what was an interplanetary toll. It is unclear how Applewhite knew the conversion rate so exactly, or how he was able to convince the group that a pile of twenty-three quarters would be transported with them and they would not be stranded at the planetary toll booth for eternity. Presumably neither economics nor physics were covered in high levels of detail during the Heaven's Gate study sessions.

Ultimately, there is little (zero) scientific evidence that Halley's comet was the cause of Mark Twain's death, or that any spectacular cometary flybys have been directly responsible for killing people beyond their own human choices. Nor is there any evidence that comets physically upset anything at all on Earth when they pass our view. However, there is abundant historical evidence to suggest that people *think* comets are heralds of doom and destruction and, as such, comets have indirectly caused a fair amount of secondary havoc. But enough about eclipses, supernovae, comets, and secondary havoc; we live in a day and age that requires nothing less than primary havoc, dammit!

Three

Humans and Heavens Collide

During human history, occasions like passing comets, distant supernovae lighting up the sky, and lunar and solar eclipses are obviously
big events witnessed by large swaths of the world. These happenings
sometimes set actions in motion that lead to lasting and important consequences for the human population living at that time, occasionally
drastically shifting the path of history. But ultimately (and respectfully),
occurrences such as comets and eclipses are really not much more than
flashing lights on a grandiose scale. And flashing lights on a grandiose
scale may get you human sacrifices, excuses for consequential battles,
and maybe even a regime change or two. Nevertheless, such flashing
lights do not leave any physical reminder to future generations of their
otherworldly nature. The arrival of a meteorite can have all the hullabaloo of flashing lights, but it also comes with the terrifying bonus of a
sonic boom and chunks of actual material falling from the sky. These
added dimensions, and maybe most important, the physical sample that
memorializes the event, have influenced humanity on a much deeper
level than flashing lights alone. Arrivals of meteorites in human times
have not just worked their way into stories and lore, but they have significantly altered, and even created religious and cultural teachings of
numerous groups of people with long-lasting consequences for billions

of individuals. Meteorites certainly are not *the* reason for religion, nor did they create human culture. Humans would have certainly developed these things whether rocks fell from space or not. But rocks from space have played a very important role in influencing human culture and religious teachings over time, and that, to me, is a story worth telling.

Quite possibly the earliest known human-meteorite contact comes courtesy of small rusted meteoritic beads found in a grave dating around 4000 B.C.E. at Tepe Sialk in modern-day Iran. The fact that these meteoritic beads were found inside a grave site provides strong evidence that they were revered samples. It is easy, then, to speculate that these adornments were part of a religious ceremony for the burial of an important person, thereby linking religion and meteorites more than six thousand years ago. But we do not know the purpose of the meteoritic beads; we do not know if the inhabitants of the area saw the stones fall and fashioned the beads out of them or if the beads were just made from an "interesting" metal rock found lying around. And we certainly do not know how/if the inhabitants of Tepe Sialk viewed them as religious objects or just cool bling to chill with during their dirt nap. Either way, the fact these materials were put into a tomb shows that inhabitants of the area appreciated the value and uniqueness of these rocks.

The study of human culture prior to chronicled text is a difficult endeavor. Before humans gained the ability to record events and the reasons behind actions of the time, context and feelings of the people involved are basically impossible to know, and we are really just throwing out educated guesses as to what was going on. However, around 3300 B.C.E. things changed when written language was invented. Hieroglyphs and cuneiform became the *fonts de force* in the earliest civilizations and humans started to record and contextualize the events of the times, providing lasting insight into human actions.

Iron in the Pre-Iron Age

About the time humans were figuring out various ways to write things down more than five thousand years ago, we were also getting fairly proficient at making bronze. Bronze is a metal alloy that represented a significant step forward regarding the production of everyday tools and warfare compared to previously used instruments/equipment made of stone and/or bones, but its production requires sources of both copper and tin, ores that are not easy to come by in most regional areas due to the rarity of tin deposits. Extensive trade filled the gap most of the time, but pervasive warring and ever-depleting sources routinely cut supplies of one or the other metal, leading to the rise or fall of various civilizations. Back in the day, basically if you could not make decent weapons or plows, your merry crew was likely not going to survive long. Every group needed a steady supply of decent metal, and if you did not have it, you were probably going to die.*

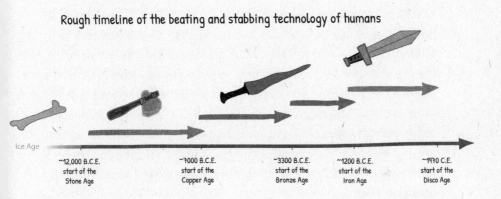

Rough timeline of the beating and stabbing technology of humans

Ice Age

~12,000 B.C.E.
start of the
Stone Age

~7000 B.C.E.
start of the
Copper Age

~3300 B.C.E.
start of the
Bronze Age

~1200 B.C.E.
start of the
Iron Age

~1970 C.E.
start of the
Disco Age

The basic tools and weapons available for mass production by humans as a function of time.

The rarity of copper, and especially of tin, in the Earth's crust presented a major problem for humans for millennia. But there was another metal, a more mystical, and far shinier metal than the drab copper or bronze that was available at the time, and that was iron metal† that came

---

* Think of it like ancient Wi-Fi.

† The word "metal" can mean many things, but here it is that iron is in its "elemen-

from meteorites. Iron may not seem like a mystical and exciting metal to residents of modern times, but five thousand years ago, humans had no idea how to make it. If you stumbled across a giant chunk of something you have never seen that could be extremely useful, you would be excited. This excitement and mysticism would grow exponentially if you saw that chunk of this shiny metal fall from the sky. Meteorites will always be awesome and somewhat mystical, but they were a whole heck of a lot more so when humans did not know where they came from and contained material humans didn't know how to make. The element iron is incredibly abundant in the Earth's crust, but it certainly does not look like it if you think of iron as just a shiny metal. The crust has virtually no metallic iron at all.* The iron in Earth's crust is locked up in regular rocks and in iron-rich ores such as Fe_2O_3 and Fe_3O_4. In order to turn iron ore into iron metal for use in things like plows and swords, you have to be able to extract it from rocks that contain it. This was far more difficult than making bronze. In short, this process requires heating the crap out of iron-containing rocks and converting iron oxides into iron metal before you can smash and pound it into whatever you want to make. However, if you found an iron meteorite, all you needed to do were the smashing and pounding parts (and maybe some mild heating). Humans showed there was no shortage of brawn and patience if the reward was an elegant silvery dagger, the likes of which most had never seen before.

Importantly, in order to produce even small amounts of low-quality iron metal from iron ore, a furnace must be able to sustain a temperature of around 1200 °C. Since early pottery kilns were only capable of temperatures of ~900 °C, higher-temperature furnaces had to be invented. In addition, producing useful iron metal requires B.C.E. removal

---

tal" form (Fe^{+0}) and not in its far more common "ox B.C.E. " form (Fe^{+2} or Fe^{+3} where it strongly bonds with oxygen to form FeO or Fe_2O_3, respectively). In Earth's crustal rocks, iron is found only in its oxide form because there is so much oxygen around.

* There are exceptions to almost anything. Minor occurrences exist in a few locations, but the only substantial deposit of native or "telluric iron" is found on the amazingly named Disko Island, off the west coast of Greenland.

of a host of impurities from the iron ore, an extra technological hurdle. Furthermore, in order to balance the hardness and strength of the final product, carbon must be admixed during the process. These capabilities remained wholly undeveloped until around 2000 B.C.E., and were extremely limited in scope and location until around 1200–1000 B.C.E. As such, the emergence of iron smelting technology around the globe is

Two methods of iron production from ancient peoples. Humans did not develop the capability to produce iron metal at all until around 2000 B.C.E., but it was not widespread until after around 1200 B.C.E.

widely used as an important historical marker by archaeologists, and is a strong indicator of not only technological advancements of a certain culture, but also of the flow of information and artifacts from travel and trade. From an archaeological standpoint, it is important that iron produced by humans and iron scavenged from iron meteorites are distinct in their compositions. Meteoritic iron contains far more nickel than man-made iron, making the types fairly easy to tell apart. So, when archaeologists find tools, trinkets, or talismans, it is possible to determine if the source of the iron is anthropogenic or meteoritic.

Ancient Egypt

The ancient Egyptians are among the earliest cultures on the planet, and meteoritic materials were in use as early as 3600–3350 B.C.E., as evidenced by prehistoric iron beads found from the Gerzeh cemetery on the western bank of the Nile. However, just like the beads at the aforementioned Tepe Sialk in Iran, the Gerzeh beads predate the written word and no contextual scribblings exist to explain how they came to find the iron or why it was stashed in bead form with a dead dude. So, in order to get any hard information about the Egyptian/meteorite connection, we basically just have to wait for writing to happen.

The traditional culture most people associate with ancient Egypt began with the unification of the north and south regions around 3150 B.C.E. This is a nice spot in time to start our discussion, since it is also around the time when Egyptians started jotting stuff down using hieroglyphics. Ancient Egyptian language, culture, and religion were incredibly dynamic and complex, but it does not take a leading Egyptologist to understand things in the sky heavily influenced their culture and religion throughout. Egypt's primary and most powerful god was Ra, the god of the Sun. But the ancient Egyptian connection to the sky certainly did not end there; Egyptians, along with most other ancient cultures, developed a deep connection to the night sky as well. This probably has something to do with the complete lack of light pollution at this point in history, or, potentially, the complete lack of Netflix.

The tombs of ancient Egypt host well-preserved evidence that no-

One sweet blade from King T's tomb. The knife is made from meteoritic iron and the sheath is made of gold. These would, even now, be considered "pretty bitchin' to own."

bility used meteorites as adornment as far back as written records go. Arguably, the most interesting is a regal dagger forged from an iron meteorite found in the famed King Tutankhamen's tomb. Alongside the dagger was a collection of meteoritic beads, rings, and other ethereal bling to journey in style to the afterlife. It is unknown whether these artifacts were witnessed falls of rocks from the sky, or simply represented a unique and natural source of iron that humankind was yet incapable of producing. To this point, around 30 nickel-rich iron objects of likely meteoritic origin have been collected from Egyptian tombs spanning approximately 2,000 years. Although not all of these objects have been confirmed meteoritic, in all cases the iron appears in tombs of high-status individuals, signaling the elevated importance of the iron objects to the inhabitants of the Nile River valley.

As if unearthing golden headdresses and meteoritic knives is not exciting enough, we can also dig into the exciting realm of forensic etymology to get a different view of the Egyptian culture-meteorite connection. From before 3100 to around 1300 B.C.E., many scholars argue that the ancient Egyptian word *bia'* (𓐊) in ancient Egyptian was, incredibly specifically, used to describe any artifacts fashioned from any meteoritic material. This would be a pretty strong indication that if all rocks were thought to be created equal in Egyptian culture, some rocks were more equal than others. However, fairly suddenly around 1300 B.C.E., the expanded phrase *bia' n pet* (𓐊-𓈖-𓊪𓏏)—which translates to *iron from the sky*—not only started to be used for meteoritic iron but was applied indiscriminately for iron from that point onward, be it man-made or meteoritic. Importantly, crude iron smelting did not exist in ancient Egypt until ~950 B.C.E. and was not widespread for another three hundred years, so what caused such a dramatic etymologic evolution in ancient Egypt hundreds of years prior to iron being produced in the area? The most likely explanation is the fall of a large iron meteorite observed by a large number of people. As was noted in a 2013 work by Diane Johnson and colleagues, it is not likely that a localized event only seen by a few people would have caused such a sudden and dramatic change in language. However, such a lexicon shift would occur if many people witnessed a meteorite shower or large iron meteorite impact, leaving no doubt that the iron that now lay on the ground had come from the sky. Conjecture? Sure, but solid conjecture bolstered by scientific observation.

So, was there a meteorite impact witnessed in ancient Egypt that provided not only an important source of iron but new vocabulary and insight into where other known pieces of iron came from? If you are looking for a figurative (and somewhat speculative) smoking gun, the 45-meter Gebel Kamil crater in southern Egypt was produced by an iron

The hieroglyphic term bia' n pet, *literally translated as "iron from the sky."*

meteorite within the last five thousand years. Sure, it may be a coincidence for an entire culture of people to almost simultaneously start referring to all types of iron as *iron from the sky* when there is evidence of a large nearby crater caused by an iron meteorite. It also may be a coincidence that NFL wide receiver Chad Johnson legally changed his name to Chad Ochocinco for a few years while he was wearing jersey number 85. But perhaps there was a connection and these things were not coincidences?

Honoring the dead with precious and unique stones is one thing, but the active worship of meteorites is another level of space rock love. These activities persist even in the modern day, but it appears that ancient Egyptians may have started the rock-worship craze with the temple of the Sun god at Heliopolis, a site of particular fervor. Located at the temple was the renowned "Benben stone," a conical-shaped rock that was displayed atop a tall pillar to receive the first light of the temple

---

* Perhaps the most recent entry is the "Church of the Meteorite," a group of 50+ believers that started to worship the meteorite Chelyabinsk after it spectacularly arrived in Russia in September 2013. The founder of the church, Andrey Breyvichko, claims that the meteorite is so powerful it could trigger the Apocalypse. True or not, the meteorite did trigger a wave of interest for people watching YouTube videos of the strange occurrences recorded by Russian dashboard cameras.

every morning.* Although the Benben stone was lost in antiquity, many scholars believe that because of its high level of importance, association with light, and conical or mound-like shape, the original stone was a meteorite.

Even with meteorites as adornment of VIPs and direct objects of worship, perhaps the most important example of meteorites in Egyptian culture was their use in the "opening of the mouth" funerary ritual, a series of pre-burial actions performed by a priest so that the deceased could eat, drink, breathe, and speak again in the afterlife. In a practical sense, this resulted in multiple broken teeth and facial bones. In the early days, priests would use a forked flint knife for the procedure, but when the power of meteorites was discovered, they enthusiastically converted to using tools made from meteorites in the ceremony. It is unclear if this "power" was learned from a witnessed fall that shook the earth or the fact that iron instruments were far stronger and more durable than anything else available, or both. But either way, meteoritic iron tools became the preferred chisel to, well, break open people's faces before they were entombed.

The Meteoritic Middle East

Along with the ancient Egyptians, one of the earliest sources of the written word came from the Akkadians of Mesopotamia. Numerous references to meteorites and fireballs can be found from as far back as 2000 B.C.E., and multiple unearthed artifacts exist that show iron meteorites were used as adornment prior to being written about. Not surprisingly, inhabitants of the Fertile Crescent treated fireballs in the sky as celestial signs; however, oddly enough, and unlike many of their cousins in other parts of the ancient (and modern) world, the Akkadians ap-

---

* The name "Benben" derives from the Egyptian word *weben,* meaning "rise, shine-forth," and is named after the sacred mound of creation where light was originally bestowed upon the world, as the story goes.

peared not to get too worked up about them and did not freak out like other peoples have throughout history. As the legendary historian Otto Neugebauer noted in 1945:

> *Mesopotamian "astrology" can be much better compared with weather predictions from phenomena observed in the skies than with astrology in the modern sense of the word.*

To the residents of Mesopotamia, fireballs, comets, meteorites, and the like were simply treated as communications from the gods, not representations of gods. References to the overhead omens were written as matter-of-fact statements and were simply foretellers of information, nothing more. If these messages were cataloged and understood, then such messages would be true for the future as well. Much like if in the past as a kid you ate twenty-six prunes as a snack and your following gymnastics routine was not particularly pleasant, you would then make a mental note that, in the future, you should *not* eat twenty-six prunes and then participate in gymnastics, at least for a day or so.

A few examples below demonstrate this straightforward and almost practical sounding approach to skyward prophecy:

> *If a train of a fireball occurs in the south and it moves along then stands and then breaks up and the daylight is scattered, the prince that has gone on a campaign will acquire all kinds of wealth.*
>
> *If a meteor flashes from the middle of the sky and disappears (on the horizon) in the west: A mighty rebellion will break out in the land.*
>
> *If, on the 14th day of the month of Du'ūzu [fourth month of their calendar, equivalent to ~June/July], an eclipse occurs and a great star falls: famine will be in the land.*

Read today, these statements may seem both a bit practical and a bit prophetic, but one may also look at it slightly differently. Realistically, any prince who ever came back from any campaign at this time in history would have very likely acquired substantial wealth, or otherwise he would not have come back. Secondly, mighty rebellions and famine were both probably more common than any major celestial theater: this is basically predicting something where the result happens far more often than the apparent cause, making the prediction far less impressive. However, as a Sagittarius, I am obviously a born skeptic about predictions from the stars.

Ancient history is littered with art, stories, and written documents that show or describe what *seem* like could have been recognized as meteorite falls, but recorders of history, at times, can be a bit hyperbolic. It can be incredibly difficult to interpret if something was literally "falling from the sky," or there was just a bit of creative license to make things seem more exciting. Interestingly, however, one very specific translation from Akkadian text seems to actually document a situation where a meteorite seems to have fallen on a man's property:

> *You take in your hand the star which fell upon your property. You recite this incantation three times. You seal it in clay. You throw the "images of well-being" into the river, and say as follows: "Lord of Heaven and Earth turn this sign away from me!"*

One could argue that this is evidence that a meteorite was known to have fallen and recovered during this era, which would make it the first direct evidence of a meteorite fall. However, you could also make the case that it was a recipe of what to do *just in case* it ever happened. When other passages on the subject are translated, the second possibility is heavily bolstered by statistics and probability.

> *If the stars fall from the sky upon the man: calumniation will befall the man. If a star falls upon the house of the man: unfounded accusations will be heard.*

Keep in mind there have only been two confirmed cases where meteorites unambiguously hit human beings in the last 200+ years, and in both cases only after the meteorite bounced off other things before hitting the person.* Meteorite-house interactions are only slightly more common in modern day, and there are exponentially more of both humans and houses now than during the time of the Akkadians. So, these statistics make it very unlikely that any Akkadian was ever hit or had their house hit with a meteorite, meaning the translated passages are more preparation for the occasion, if it should occur. Either way, if there were any meteorites collected in ancient Mesopotamia, they are almost certainly currently residing at the bottom of a river.

Other neighboring cultures, such as the ancient Anatolians (modern-day Turkey), had less contentious relationships with meteorites. In fact, according to many ancient texts starting around 1900 B.C.E., a precious metal called *amūtu* developed a value around eight times that of gold and was noted for its scarcity in markets. *Amūtu* seems to have been primarily ornamental in use, and while *amūtu* cannot unequivocally be pegged as meteoritic iron, there is certainly a case to be made that it was. Sure, it is *possible* that a small group of secretive and technologically advanced people were setting up small-scale, clandestine iron smelting operations intermittently around Anatolia for hundreds of years, smelting iron here and there and not telling anyone how, keeping the price of *amūtu* high (and effectively covering their tracks to this point). But it is not likely. It is far more likely that an industrious group of people found one or more large iron meteorites and painstakingly chipped off pieces—using the existing and effective technology of banging repeatedly—and made interesting things from the iron bits. When they were tired of hacking at the giant lump of iron or needed an injection of cash, they would assemble the pieces, raw or worked into a trinket, and take them to market. Unfortunately, very few examples of such ancient items exist, but the iron dagger found in a grave at Alaca Höyük is a notable exception. The tomb in which the dagger was found

---

* A large wooden radio, and a banana tree in case you are curious what they bounced off. And there is zero chance you could have guessed those, so don't even act like you would have.

A dagger from Alaca Höyük, Turkey, likely made from a meteorite sometime around 2500 B.C.E. The serrated blade is clearly made for slicing bread extra thin.

dates to around 2500 B.C.E., predating widespread knowledge of iron-making by ~1,300 years. Preliminary work has shown that the high nickel content of the blade suggests a meteoritic source, but this has yet to be confirmed. Additionally, my independent research from high-tech image analysis shows that I have no interest in being stabbed or cut by that jagged blade, and this would have been especially true prior to modern wound care and antibiotics.

The people inhabiting Anatolia changed from one culture to the next over the centuries of warring and migration, and many recorded their eventful histories through the various writings of each group. And let's face it, most ancient writings say nothing about meteorites, and even when they do, they can be quite vague when it comes to details about how they were used. However, this is not always true. One passage in particular from Hittite ritual writings about how to build a proper house, of all places, discusses meteorites quite specifically.

The diorite they brought from the earth. The black iron of heaven they brought from heaven. Copper (and) bronze they brought from Mount Taggata in Alasiya (Cyprus).

Sure, the building materials for this house they are describing may seem a bit opulent and a bit out of reach for the common dirt farmer living in a straw and mud hut, but this passage also importantly shows that the Hittites of 1500 B.C.E. were hip to the primary source of iron at the time. Iron came from the sky, not from Earth itself. For me, it is quite interesting to think that around 3,500 years ago, entire cultures were fully aware of the fact that things could fall from the sky and not be from this planet. Yet, at the same time it is depressing this knowledge was somehow lost or ignored. The general consensus from "learned folk" as recent as the 1800s claimed that there was no way anything of the sort could happen, and that reports of the like were only nonsensical commoners begging for attention.

Ancient China

Chinese cultures have existed in vast areas of East/Southeast Asia for well over five thousand years. And throughout history, the Chinese have proven to be excellent recorders of events, particularly those of an astronomical bent. In fact, the earliest clear reference to a meteorite fall* was recorded in 2133 B.C.E. in Xiaxian, Shanxi Province. Over the following ~4,000 years, historical documents note >350 instances where meteorites are witnessed to have fallen and the stones recovered in various Chinese provinces. With all of this thorough documentation and apparent sky rock collecting, one would reasonably predict there is an extensive collection of ancient meteorite falls in China and we would fully understand how the ancient Chinese viewed rocks falling from the sky.

From a physical object perspective, we know from well-dated artifacts that the Chinese used meteoritic iron when forging special weapons for special people prior to both record keeping or their ability to smelt iron.[†] The earliest known such artifact is a bronze axe with a meteoritic iron blade dated to the fourteenth century B.C.E. (Shang Dynasty), but better-preserved examples of similar objects exist from the early Chou Dynasty (around 1000 B.C.E.). Unfortunately, no records exist detailing if these workable slabs of iron fell in the area or were acquired in trades;

---

* The difference between a *fall* and a *find* in meteoritics is important. In the first, you see it fall from the sky, the second you just find it on the ground. Clever naming scheme, eh?

† Written records in China started to appear around 1250 B.C.E. Iron was not smelted routinely in Chinese culture until around the fifth century B.C.E.

Chinese (Chou Dynasty) dagger and axe-head weapons. These are made from a mix of bronze castings and meteoritic iron blades. The iron parts (present and former) are shown outlined in dotted gray lines.

however, based on analysis of their high nickel content, it is clear that they are of meteoritic origin.

But even though China has the ideal factors to potentially have amassed an immense historical meteorite collection—a lot of people spread over a vast area for a ridiculously long time—it simply does not host an abundance of actual meteoritic material. This remains the case even today: The Meteoritical Society's database of meteorites[*] lists only around 250 samples originating in China, with about two-thirds of these recovered in meteorite hunting expeditions in the last twenty-five years. Contrasted with the contiguous United States, the host of

---

[*] By far, the world's most comprehensive database of meteoritic falls and finds (https://www.lpi.usra.edu/meteor/).

more than 2,000 meteorite falls and finds, there is a clear discordance. This is especially evident when you consider that the lower forty-eight states are almost identical in size to China, yet have a much smaller population with a laughably shorter history. Even the far less populated (and slightly smaller) country of Australia has over 700 meteorites to its name.

Meteorites do not disproportionately hit (or not hit) the Earth in certain areas, and they certainly do not prefer landing inside any particular political boundary over another—they land in random locations around the planet. So, since there is no natural scientific reason for the dearth of meteorites from China, why does this vast difference in historical abundance exist? It has been suggested, quite seriously, that China has so few meteorites because, historically, the inhabitants are eating meteorites; soon after a meteorite would fall, it was quickly recovered by the local population and ingested in some creative way, as people thought it would cure ailments. The earliest unambiguous observed Chinese meteorite fall we have a sample of is the meteorite Jianshi, which fell in 1890. Fortuitously for meteorite researchers, Jianshi is a 600 kg iron meteorite, making it difficult to displace or digest. Such meteorite munchies may seem strange, but it is not only ancient rural China that appears to have such an insatiable appetite for the bolide buffet. In 1886, the villagers around Novo Urei, Russia, were treated to the violent fall of what became the namesake of the "ureilites," a graphite and nanodiamond-packed unique type of meteorite. However, it seems that a significant portion of the >2 kg meteorite ended up inside villagers who thought they would gain powers of the gods by ingesting a rock that came from above.* Even more recently, a shower of stones of

---

* It may not come as a surprise, but for the record I want to note that there are no documented health benefits to eating meteorites, only health concerns. If you eat a small amount, say a small spoonful of an ordinary chondrite, for example, this is probably not a big deal. However, ordinary chondrite meteorites contain ~10 percent iron and acute doses of iron can cause intestinal damage and other complications. Perhaps even less desirable are the high levels of toxic metals like mercury or cadmium found in meteorites. So maybe don't eat them; they are rocks from space, not food or medicine.

various sizes fell near the city of Mbale, Uganda, in 1992.[*] At the time, Mbale was in the midst of the AIDS epidemic and the desperate villagers thought this could possibly be a God-sent cure. As such, many pieces of the meteorite were ground into a fine powder and mixed with liquid to form a paste, which was either drank or applied to the skin. People believe what they want to believe.[†]

Aboriginal Australia

Even though there is no written language in Aboriginal culture, their oral histories date back tens of thousands of years, well before anyone was writing hieroglyphics or retweeting cat videos. Aboriginal people spanned huge regions of Australia's barren deserts for an incredible amount of time, have a deep connection to their natural surroundings, and, as a result, are observant of natural phenomenon. As such, the chance of them finding rocks that do not belong, witnessing large meteorite impacts, and passing that information along for generations is quite high.

However, despite the wealth of meteorites sitting around the Outback,[‡] there is very limited evidence that the Aboriginal people ever utilized meteorites for tools or adornment. This could partially be due to the limited technology Indigenous peoples possessed that would have enabled them to turn hard lumps of iron metal into useful tools, but it appears more likely it was a superstitious view of all things meteoritic. Aboriginal people have regarded meteoritic masses and/or their impact sites (real or perceived) with awe and fear for millennia, as things to be avoided. Is there a reason for the almost ubiquitous distrust of all things meteoritic from the Aboriginal population? The answer to this

------

[*] This is one of the stones mentioned earlier that struck a child after bouncing off a banana tree.

[†] Don't get me started on similar ideas about ingesting endangered species.

[‡] Western Australia and, in particular, the Nullarbor Region is one of the most productive meteorite hunting grounds on Earth. This is largely because most meteorites are dark, and the Nullarbor Region country rock is flat, pale limestone, so alien rocks are quite recognizable.

understandable stance might actually come from personal experiences of the Aboriginal population with stuff falling from the sky. Multiple craters found in Australia are young enough to have been seen by the Indigenous populations, and in at least one case, the Henbury crater field, it seems that the event was witnessed by the local population, possibly leading to warnings to hundreds of generations to come about the dangers of anything meteoritic.

The Henbury crater field was formed after a meteoroid broke apart in the atmosphere, raining down multiple pieces of meteoritic iron over a square kilometer, creating a series of thirteen craters. The craters have been dated to be around 4,700 years old, so well within the time frame of Aboriginal people living in the area. Evidence for human witnesses to the cratering events comes from oral traditions collected in the 1930s, in which Henbury is referred to as a place where "a fire-devil came from the sun and ran down to the Earth, creating the craters." It is said that the spirit "burned and ate the people for breaking sacred laws," and Aboriginal people still will not collect water from the craters "for fear the fire devil would fill them with iron." Still to this day, many Aboriginal people will not camp within even a few miles of the craters and refer to them as *chindu china waru chingi yabu,* roughly translating to "sun walk fire devil rock," which seems like a reasonable name if your ancestors saw something flying out of the sky that destroyed an entire area.

The Aboriginal wariness of rocks from the sky and the damage they can cause appears to have transferred to many geologic features similar to Henbury, such as the Wolfe Creek crater, more than one thousand kilometers away from Henbury. The Wolfe Creek crater was formed around 120,000 years ago, meaning that it is very unlikely any humans saw it happen. Yet, formation of the Wolfe Creek crater is dramatically described in the oral history of the area:

> They saw a big bright light come down from the sky, coming down like a ball of fire. It shook the ground. The people hid in a cave, because the ground shook real strong. They saw dust coming up from the ground. When that settled down they were talking to each other. They didn't

Above: *Aerial picture of the Wolfe Creek crater.* Below: *Aboriginal art rendering of the area and how it formed.*

want to go closer where that thing fell. They never touched that area.
The star that fell down was evil. It was an evil thing. [Artist's story,
Jane Gordon, Billiluna—https://web.sas.upenn.edu/psanday/exhibition/
painting-gallery/]

In Aboriginal lore, the Rainbow Serpent is among the most famous of the creator beings, giving form and order to the land and society. It is described to move like a "snake that travels like stars travel in the sky." Again, descriptions like these are difficult to parse concrete, physical meaning from, but it sure seems like the Aborigines were witnesses to at least some impressive cosmic phenomena that greatly shaped not only their homeland, but also their spiritual and cultural views.

You may think some of these described meteoritic connections to metaphysical beliefs/experiences are explained due to their distance in time and culture from your own (whatever it may be), but these connections are not just confined to a handful of people or related to something thousands of years ago. Meteoritic influences in religion and society are pervasive in almost every corner of the world, ranging from modern Oceania, Indigenous peoples of the Americas, to Estonian folklore—regardless of where on culture's timeline you look.

Creeping up to Modernity

Around 1000 B.C.E. in what is now modern-day Iran, the fabled and enigmatic proto-Persian prophet Zoroaster was many things by many names,* but one of his persistent legends was that he was a man who was knowledgeable about the stars, and he used that to his advantage. Even though firm details about the legendary man are rare, it appears Zoroaster noticed the periodicity of, and subsequently predicted, meteor showers such as the Leonid meteor shower. These predictions eventually netted him a large following of zealots impressed by the falling star

---

* Zoroaster is sometimes known as Zarathustra, as well as various other "Z" or "S" names.

Image of Zoroaster with a glint in his eye, and possibly his meteoritic weapon used to smite demons.

forecasts and are likely what earned him his name, which translates into "the living star."

Zoroaster may be unknown to many modern people, but his contributions to modern culture are ubiquitous. Perhaps the most notable contribution was the introduction of dualism. Zoroaster's dualism dealt with the clash of good and evil, and somewhat paradoxically, dualism is widely regarded as the basis for humankind's first monotheistic religion. If that was not enough, his dichotomous dictations developed the ideas of Heaven and Hell (not the Black Sabbath album), angels and demons (not the movie), as well as the concepts of Judgment Day (again, not the movie) and Satan (yeah, that one). Finally, his fundamental principle of the struggle between goodness and light, and darkness and evil, has some pretty easy-to-draw parallels with Star Wars (of course that one).

Zoroaster's connection with space was not just in the cosmic fireworks he predicted to gain a following; he had a more tangible association to the heavens in what was rumored to be his favored weapon. Zoroaster is said to have routinely defeated various demons with a "massive stone received from God,'"* which certainly sounds an awful lot like a meteorite. And given that Zoroaster lived prior to iron production in the area, a blade made of meteoritic iron would be far superior to any other metal weapons available, particularly against demons, which likely also lacked the ability to smelt iron. Zoroaster became a legend thanks to a variety of space rocks, and if he had never become a legend, he may never have laid the foundations of Judaism, Christianity, Islam, or Star Wars. A sizable legacy.

Greeks and Romans

The ancient Greeks and Romans had an impressive run in Europe and beyond, spanning well over a thousand years, and there are an impressive amount of both Greek and Roman records discussing objects that were seen to fall from the sky—far too many to discuss at any length here.† To summarize the literature of the time, it appears all of these fallen objects were treated with reverence as tokens from the gods. As these were such important objects, some of the larger and more important ones received impressive shrines built in their honor, and some meteoritic arrivals were even commemorated with the pressing of coins depicting the event. The celebrated Stone of Aegospotami, a meteorite "the size of a wagon load" that fell around 465 B.C.E., was discussed at length by notable names like Pliny the Elder, Plutarch, and Aristotle, and the shrine built in its honor was a local landmark for more than five hundred years. Another example is the famed statue of Diana, originally housed at the Temple of Artemis (one of the seven wonders of the ancient world). The statue was widely reported to have been made of meteoritic material and is quoted in numerous sources as having "fallen from heaven." Presumably, the 2.9-meter-high statue was carved from a rock that fell to Earth, and the statue did not fall pre-carved,

---

* Hokey religions and ancient weapons are no match for a good meteorite at your side, kid.

† An impressive compilation can be found in D'Orazio (2007).

but again, translations and interpretations can be occasionally cumbersome. Sadly, neither the Diana statue nor the temple itself survived the ravages of man and his propensity to pillage and plunder. In fact, none of the meteorites of these ancient times appear to have survived to modern day thanks to a variety of wars, fires, and possibly even some brimstone.

Nevertheless, one meteorite-related nugget that has survived the test of time from this era is the historical note that the Roman Empire, arguably one of the greatest civilizations known to humankind, officially worshipped a meteorite for the four years spanning 218 to 222. How this happened is a winding, complicated, strange, and bloody story that would fit perfectly with numerous characters' attempts to rule Westeros in *Game of Thrones* (complete with multiple assassinations of emperors, bastard-son claims, overpromoting grandmothers, and lurid sexual adventures). The CliffsNotes version of the tale starts with a witnessed meteorite fall in the desert surrounding the town of Emesa (modern-day Homs in Syria). A small nomadic group of Sun-worshippers obtained the stone, thinking it was the embodiment of their god *Ilâh hag-Gabal* (whom the Romans called the only slightly more pronounceable *Elagabalus*). Now that they had a proper idol from the heavens, the nomads built a temple and began a-worshippin'. It just so happens that one of the fast climbers on the career ladder of high priests of this tribe was identified, by his ambitious grandmother, as the bastard child of a prior emperor Caracalla. Without any other clear options and the need for a head of government, this fourteen-year-old rock-worshipping zealot was proclaimed Emperor Elagabalus. To put it mildly, the newly minted emperor was not the typical Roman ruler, if there ever was such a thing. Elagabalus started things off with a bang, an almost year-long elaborate celebratory transfer of the stone he worshipped to his new city of Rome. Upon the stone's arrival, and what he thought was a sign of respect for the religion of Rome, Elagabalus publicly (and quite extravagantly) married the "male" meteorite to multiple "female" tributes of the Roman religion. In what must have been a very bizarre thing to watch, the wedding tributes were all adorned with gems and fine trimmings, complete with the utmost pomp and circumstance. Following the ceremony, the Stone of Emesa became an officially worshipped god of Rome, replacing Jupiter as its most important religious figure.

The remaining short tenure of Elagabalus was no less interesting and, let's say groundbreaking, for the Roman Empire. In a progressive milestone, Elagabalus's reign saw the first women allowed into the Senate, although it seems a tickle less progressive when they are your mother and your grandmother. Elagabalus broke down plenty of other boundaries when he often wore makeup and dressed as a woman, and reportedly routinely prostituted himself in local establishments of ill repute. At one point, he allegedly demanded a sex-change operation, only to be greatly disappointed when surgeons of the time told him they did not possess such abilities. Possibly angered by this, or wanting to seem manly to the Roman public, he raped (by some accounts), then married (by all accounts) a vestal virgin* named Julia Aquila Severa, with promises that this would create "god-like children." This was his second of what would be five marriages in his short time as emperor. Eventually, the eccentric and often quite horrible actions of Elagabalus became too much for the power brokers of Rome and this meteorite-worshipping hiccup ended in the fanatical ruler being dismembered and tossed into the Tiber River.† An end George R. R. Martin would certainly approve of, no doubt.

Cosmic Connections in Christianity

The era of Jesus Christ in human history not only coincides with the calendar switching from awkward negative numbers to counting years in a positive direction, it also set in motion a few other notable changes.

---

* According to Roman law and tradition at that time, any Vestal found to have engaged in sexual intercourse was to be buried alive. Regardless of whether Elagabalus forced himself on her or not, he clearly did not have her best interests at heart.

† If you are curious about the fate of the Stone of Emesa following the unseating of Elagabalus, it was unceremoniously returned to the temple previously built for it in Syria. However, the meteorite almost certainly was destroyed and mixed with earthly dust when rowdy Christians in a combative mood descended on the cult's temple around a century later.

Above: *Reconstructed miniature of the Temple of Artemis that housed the statue of Diana.* Below: *A Roman coin minted in honor of the sacred Stone of Emesa. Not to scale.*

And some of them very much appear to be rooted in cosmic phenomena. Earlier, we discussed the possible cometary connection to the birth of Jesus Christ. Since the events of, and writings about the life of Jesus have turned out to be fairly important to humanity, let's spend a bit of time on the meteoritic musings of the authors of various religious texts in Christianity.

Of course a fundamental concern when investigating ancient writings about meteoritic origin, or really anything else, is historical reliability. In the study of ancient meteoritics, this is compounded by the fact that meteorites come from the direction of the sky (aka "the heavens," or "above"), the precise direction most deities were thought to reside. Therefore, it is very easy to imagine that if a resident, in fact almost any resident of any culture 1,000+ years ago, found an interesting chunk of rock that looked supercool, it is expected they might say that it is "from the heavens." Even in modern times, it is common to hear someone say that babies and particularly good-looking people are "gifts from above," so you can hopefully excuse the difficulty for scholars in translating ancient text for literal meanings in a few confusing terms. This difficulty exists in all ancient texts, but it particularly thrives in early Christian writings. For one example, the book of Revelation in the Christian Bible states in chapter 6, verse 13:

And the stars of the heaven fell to the earth, like unripe figs dropping from a tree shaken by a great wind.

It is certainly possible this verse is describing the falling of meteoritic material—the first part very much sounds like it could be—but unless the fig trees 2,000+ years ago were substantially taller than they are now and the winds were much more violent, that description represents a bit of an understatement about how meteorites actually fall to Earth. Consider it noted that terms like "understatement" are not generally consistent with the Bible, a book stuffed with countless tales and/or metaphors that not only violate common sense, but quite routinely, the laws of physics. As such, we end up looking through documents that are filled with hyperbole and confusing language. While these literary devices make these documents a lot more exciting to read, it does not always make them easy to interpret from a scientific perspective. In fact, subsequent parts of the book of Revelation can be read like a pretty intense meteor shower described instead by the author as seven angels blowing seven trumpets and making a serious mess of things:

> *The first sounded, and there came hail and fire, mixed with blood, and they were thrown to the earth; and a third of the earth was burned up, and a third of the trees were burned up, and all the green grass was burned up.*
>
> *The second angel sounded, and something like a great mountain burning with fire was thrown into the sea; and a third of the sea became blood, and a third of the creatures which were in the sea and had life, died; and a third of the ships were destroyed.*
>
> *The third angel sounded, and a great star fell from heaven, burning like a torch, and it fell on a third of the rivers and on the springs of waters. . . .*

At this point in the story, the people in the area were probably eager for the angels to put away their trumpets. It sure sounds like the events being described could be a large and destructive meteor shower, but we do not know if these verses were based on any historical events or just artistic license from the author who really hated trumpet music.

In addition to references to likely meteor showers, other meteoritic activities seem to play an important role in the Bible as well. The tale of Sodom and Gomorrah—two ancient cities mentioned in religious texts of multiple faiths—alleges these towns were destroyed swiftly due to some sort of deemed wickedness.* And while smiting a couple of cities may just feel like normal flow in the eventful Old Testament, looking at

---

* Traditionally, the story of the destruction of these cities has thought to revolve around punishment for sexual acts, often interpreted as homosexuality. I here note that meteoritic derived destruction is random with no agenda or preference for sexual orientation or behavior—tales told otherwise are likely spun to benefit the storyteller.

the story through a meteoritic lens may shed some light. The book of Genesis (19:24–25) reads:

> *Then the Lord rained down burning sulfur on Sodom and Gomorrah— from the Lord out of the heavens. Thus he overthrew those cities and the entire plain, destroying all those living in the cities—and also the vegetation in the land.*

If Sodom and Gomorrah were real cities destroyed as quickly and efficiently as suggested by multiple religions, the cause of the destruction would either be via (1) a vengeful deity, or (2) from a natural event. Discounting the first simply because vengeful deities are not the focus of this book, one type of natural event that could level multiple cities and the surrounding vegetation with the efficiency of a modern nuclear warhead is something known as a meteoritic airburst. These impressive explosions have been documented in modern times and occur when a meteor explodes with great force as it makes its way through the atmosphere.*

Whereas the idea of a meteoritic airburst destroying ancient cities is circumstantial at best, one of the connections between Christianity and the cosmos that we can look at from a more scientific perspective is the potential connection between a significantly smaller meteoritic airburst and story of a man who would become one of the most significant evangelists in early Christianity. After the crucifixion of Jesus, Christianity did not immediately become a widespread religion overnight with millions upon millions of devoted followers; it remained an obscure Jewish sect for quite some time. It needed advocates and enthusiasts to get the ball rolling. Arguably the most important of these first

---

* Certain researchers have suggested that the archaeological site of Tall el-Hammam outside the capital of Jordan contains physical evidence of an extremely high temperature event and have suggested that Tall el-Hammam is the destroyed ancient city of Sodom. At present, this does not appear to be a wildly held view in the scientific community, but perhaps the ongoing excavation of the city will reveal compelling evidence in the future.

millennium influencers was a guy named Saul who, at first, was one of Christianity's biggest naysayers and frequent persecutors of early Christians. As the New Testament story goes, Saul was traveling with his group of followers in search of disciples of Jesus to rough up. While traveling on the road to Damascus, Saul and his gang witnessed a dramatic flash and boom that knocked Saul over and blinded him for a few days.* Apparently, this extreme experience changed his views in an instant and Saul decided to switch teams. He promptly exchanged the "S" for a "P" (some changes are more subtle than others) and the new man Paul the Apostle took up missionary journeys across the Mediterranean. Paul went on to become not only the namesake of the capital of the state of Minnesota, but also the principal architect of the spread of Christianity around the Mediterranean, and arguably the most important evangelist in Christianity— perhaps all due to a meteorite airburst on the road to Damascus.

Buddhism?

Because Buddhists do not worship gods or deities, but instead strive for a deep insight into the true nature of life, you might think it would be difficult to find a meteoritic connection here. But you would be mistaken. In Tibet, meteoritic iron has long been known as *namchang*. This translates to "sky iron," suggesting the local population, even long ago, was aware of the skyward origins of at least iron meteorites. Again, Western society really dropped the ball on that nugget of info for quite some time. While this meteoritic connection does not seem to have shaped the Buddhist religion in any notable way, an eleventh-century carving of Vaiśravana, an important figure in Buddhism, was found to

---

* There is a fantastic comparison of this biblical story with firsthand reports of what happened to some people following the Chelyabinsk airburst of a meteorite in Russia in 2013. Following the flash of light, Paul allegedly suffered from short-term blindness, skin damage, and eventual flaking of "something like scales" falling away from his eyes, as recorded in religious and historical texts. To quote from Hartmann (2015), "This striking phrase beautifully matches severe photokeratitis, with epithelial desquamation. This match is one of the strongest lines of evidence that the first-century accounts are reporting, as best they can, real phenomena."

Painting of the bright light that blinded S(P)aul and started his conversion to spread the word of Jesus Christ. Replace the angelic dude carrying the cross in the clouds with a meteorite flash, and this might have actually happened.

be carved from an iron meteorite, creating an interesting intersection of religious art and meteoritics. Work by Buchner and colleagues in 2012 identified the specific meteorite the statue was carved from, the Chinga meteorite that fell on the border area between Siberia and Mongolia more than ten thousand years ago.

Meteorites and Muslims

Perhaps the most publicized of the meteoritic/religious connections belongs to the Islamic faith. The most sacred site in Islam is the Kaaba, the direction which Muslims are expected to face during prayer. Inside the Kaaba is the Black Stone, the most venerated rock on Earth, which is said to have been set into the wall of the Kaaba by the prophet Muhammad himself in 605 C.E.. Whereas the stone had been honored in pre-Islamic times, Islamic tradition tells that it had fallen "from heaven" to guide Adam and Eve to a location to build an altar. This wording points at a meteoritic origin, but ever-present complications from language and meaning prevent any scientifically definitive provenance of the rock on this alone.

An eleventh-century carving of Vaiśravana made from the iron meteorite Chinga.

By all accounts, the Black Stone has been through a lot: it has been burned, stolen, smashed into fragments, survived floods, and polished smooth by the hands of millions of pilgrims. It was reportedly originally a single stone, but it now appears to exist as eight pieces cemented together and protected by a silver frame. Multiple reports over the centuries provide informative, yet conflicting, evidence on whether the Black Stone could be a meteorite or not. The fact that it has been broken multiple times essentially rules out that it is an iron meteorite, at least. One report from the year 951 suggested the Black Stone was recognized as an original following its theft in 930 because it was known to float in water. If this report is correct, the Black Stone would not be meteoritic, since none of the known meteorites are less dense than water, and instead point its provenance toward some sort of volcanic pumice. The original color of the stone is another source of uncertainty since most meteorites are dark in color, and some reports say it was as "white as milk." Of course, the original color could have been altered to black over centuries of human handling, or alternatively, due to the sins of humankind,

The Kaaba in Mecca (above), *home of the venerated Black Stone* (below) *that is found as a cornerstone of the structure.*

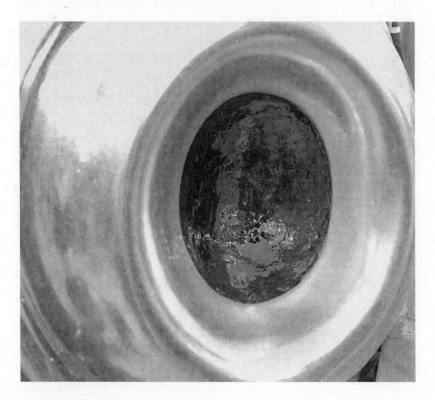

as the legend goes. Either way, even if the rock were not meteoritic, a "heavenly" connection cannot be ruled out, as discussed in work by Elsebeth Thomsen in 1980. About 1,100 kilometers east of Mecca is an impact crater in Wabar that formed around six thousand years ago, well within human occupation of the area, meaning it is possible the impact was witnessed. The local rocks in the Wabar region are of pure white sandstone that, due to the impact, has formed pumice-like blocks that can have a glassy appearance, can be white, and importantly, can contain gas bubbles allowing them to float, all consistent with the various reports of the Black Stone.

However, until scientific testing can be performed on a small piece of the stone to determine its provenance, whether or not the Black Stone is a meteorite, ejecta from a meteorite impact, or simply a terrestrial rock that garners a lot of attention will remain speculation. And while many researchers like myself would absolutely love to understand more about a possible connection of the Black Stone to meteoritic interactions in the ancient world, I am not going to volunteer to chip off a corner of one of the most sacred Islamic artifacts without permission.

Survival of Meteoritic Material

While not all religions and cultures seem to have an obvious meteoritic connection, many do. Multiple native tribes of the Americas made use of iron meteorites they came across, and some tribes held sacred the meteorites they encountered. People of the Hopewell culture are known to have traded meteoritic material around the plains of North America, and used them in burial rites. The Inuit of Greenland shaped bits of the ~35 ton meteorite Cape York into metal tools and harpoon tips for centuries. We have discussed only a handful of the hundreds of meteoritic-related stories present in various texts across geography and time. Every culture that has recorded such events appears to grasp the rarity and importance of such rocks from space. Yet, it may come as a surprise that of all the witnessed falls in human history, the oldest recorded fall that we currently have a sample of is the Nogata meteorite that fell in May 861 c.e. in southern Japan. When you look at it from a scale perspective, we add over 40,000 tons of extraterrestrial material to Earth every year, and whereas the majority of that is in the form of dust-size particles, there are still plenty of reasonably sized chunks that make it to our planet's surface. However, compounding the issue of being able to verify a meteoritic origin of described events, reported fallen material just simply very rarely survives the passage of time once humans are involved. It may have survived 4.5 billion years floating around space, but give a meteorite a few days/years in the presence of a group of human beings and you never know what will happen.

From a historical and cultural perspective, rare cosmic events appear to rank as high as anything that shaped modern society into the form it is currently in. How would the current monotheistic religions be different if it were not for meteorites? Would Europe look completely different if it were not for timely arrivals of cosmic visitors? Would you please stop eating things that should not be eaten? It can be a fun philosophical exercise to discuss how comets and/or meteorite falls have shaped the course of ancient history, but that exercise becomes somewhat more possible when saving records of meteoritic occurrences and, maybe more important, the physical meteorites themselves, became more commonplace.

Four

Prognostication, Panic, and Scientific Progress

Early writings make a good case, or at least a solidly arguable case, that ancient cultures such as the Egyptians, Hittites, Anatolians, and Chinese were reasonably aware that the source of rocks falling from the sky was, well, the sky. People of the ancient world did not seem to have a great deal of difficulty believing that rocks could originate from outside the confines of Mother Earth and yet still land on her. Sure, sometimes meteorites were mistakenly thought to be messages from demons or gods, or in the case of the more modern Aztecs, meteorites were believed to be divine excrements. And while thinking about my chosen field of study as godly-poopology makes me chuckle a bit, at least the Aztecs correctly deduced the hypersonic turds from the sky were of extraterrestrial origin and not just fast-moving, locally sourced turds somehow blowing around the atmosphere.

The same cannot be said for "learned academia" for the better part of the last two millennia, and the reason for this massive misunderstanding, at least for me, is quite surprising. One might think that it was primarily a religious reason falling rocks were dismissed as coming from somewhere beyond Earth, but it was almost the opposite of that.

In most academic circles, the belief that rocks fell from outer space was held back by an overreliance on a few brilliant scientific minds, starting at the height of classical Greek civilization. This conviction of an earthly source of falling stones, again oddly, was compounded by the even more brilliant (and even more scientific) minds yet to come.

The Grecian Contributions

Few people would choose to argue, at least legitimately, that the ancient Greeks were not a heady bunch. They spearheaded an incredible number of things: geometry, philosophy, democracy, and, of course, the Olympic Games, just to scratch the surface. In addition to their outsize contributions in most fields, the Greeks played a very important role in the history of meteoritics. However, their legacy from a meteoritic perspective was considerably less helpful than it was in most other areas of science and the humanities. Early on, things were looking just dandy for what would become the study of meteorites when Diogenes of Apollonia* reported the fall of a brown stone at Aegospotami around 465 B.C.E. and was clearly hip to its cosmic origin:

Meteors are invisible stars that die out, like the fiery stone that fell to Earth near the Aegos Potamos.

Diogenes later wrote:

With the visible stars revolve stones which are invisible, and for that reason nameless. They often fall on the ground and are extinguished, like the stone star that came down on fire at Aegospotami.

---

* Diogenes is honored for his early understanding and writings about the cosmic origin of this rock by having a type of meteorite, the *diogenites,* named after him. Of course, we don't know if this alleged meteorite at Aegospotami was what we now know as a diogenite (since humans plundered and destroyed it before it could be properly studied) but it is a nice gesture anyway, I think.

Not bad for a man living ~2,000 years prior to humans realizing that Earth was *not* the center of the Universe. However, Diogenes's ideas about the origins of rocks from the sky were not to last long in Greek academic circles, unfortunately.

It is difficult to point to a single person who destroyed the idea of meteorites coming from beyond Earth, holding back the study of our Solar System's origins for roughly two millennia, but I will do it anyway. It was Aristotle. I really hate to bag on someone who has exerted such incredible (and mainly positive) influence on almost every form of knowledge in Western culture, I really do, but misconceptions about the source of meteorites can be traced fairly directly back to Aristotle's tome *Meteorologica* in the fourth century B.C.E. *Meteorologica* is the oldest comprehensive treatise on what we now know as meteorology, or the study of atmosphere and weather. But if you stretch . . . really stretch . . . and put on your forensic etymology hat yet again, you may notice slight similarities in the words *meteor*, *meteorite*, and *meteorology*. It is subtle, but it is there, trust me.

Aristotle. A really smart fella who unwittingly really screwed us meteoriticists.

Essentially, every conceivable terrestrial and atmospheric phenomenon was addressed in *Meteorologica*, all somehow related back to interactions with the classical four elements: earth, water, air, and fire,* along with Aristotle's added divine fifth substance to complete the special sauce, the "aether." Regarding the topic of *this* book, however, the Big A thought the falling rocks and fireballs reported by his predecessors and contemporaries were either: (1) lifted into the air by great winds—as he specifically suggested for the aforementioned Stone of Aegospotami—or (2) a combination of various exhalations from Earth that, when mixed under the right conditions, materialized in the sublunary region of the Earth-Moon system. Why would one of these two possibilities be more likely than meteorites just coming from somewhere else in the Solar System, you may ask? Chiefly, it seems that an external source of meteorites presented a bit of a, well, flying rock in the ointment for Aristotle's grand vision of the Universe. Aristotle believed that beyond Earth and the Moon was the region of the stars, an area of timeless perfection that was unchanging and unchangeable.† Sure, this sounds romantic, but, like most things romantic sounding, it is completely untrue. And since the basis for Aristotle's grand vision of the Universe did not allow for anything dynamic beyond the Moon's orbit, all of those damn falling rocks must be coming from Earth itself, somehow.

Had Aristotle not been such an intellectual giant, his missteps on meteoritics would have been easier to ignore or replace by subsequent minds on the subject. But he was Aristotle, so going against him was, well, not logical.‡ Incredibly, it took until the scientific revolution of the sixteenth and seventeenth centuries, thanks in large part to the inven-

---

* Aristotle preferred his take of mixes of *dry, moist, hot,* and *cold* to the classical element names, but this gets confusing very fast and is not of much relevance here.

† This idea may sound very strange now, but considering he lived almost two thousand years before the first telescope, I am willing to cut him some slack for his astronomical inaccuracies.

‡ Aristotle is also credited with creating the first formal study of logic. It was difficult to get away from his mind-shadow for a long, long time.

tion of the telescope, for Aristotelian ideas of astronomy and physics to start to be supplanted. But even with humanity's vastly improved knowledge about the planets and their orbital relationship to the Sun, rocks falling from the sky continued to be thought of as either forming on the ground and tossed about by strong winds, or sourced from earthly exhalations such as volcanic eruptions. Unfortunately, a millennia-long succession of academic A-listers provided credence to the idea that falling rocks could be sourced from terrestrial materials. Leading up to and during the Enlightenment, René Descartes and fellow Frenchman Antoine-Laurent de Lavoisier both argued (about a hundred years apart) that lightning and atmospheric dust could be the precursors to stones that then fell to Earth. And, since contemporary electrical experimentations (largely pioneered by the American polymath Benjamin Franklin) showed that iron oxide could be converted into iron metal, some scientific squinting, hand waving, and lightning could of course explain why chunks of iron metal were repeatedly seen falling from the sky. However, the most important continuation of Aristotle's erroneous notion of Earth-sourced falling rocks came from an unlikely source near the close of the seventeenth century, someone with perhaps the most impressive scientific chops in history . . . the peerless Sir Isaac Newton.

In what is arguably the most influential book in scientific history, *Philosophiæ Naturalis Principia Mathematica,* Newton established the foundation of classical mechanics, explained universal gravitation, and derived the laws of planetary motion. Not bad. In his second book, *Opticks,* also one of the great works of science history, Newton unfortunately (and incorrectly) describes how meteorites may be formed by material rising in dry air and mixing with other materials high in the atmosphere. Perhaps even more damaging to the study of meteoritics than Newton's inaccurate ideas on how meteorites were produced, Newton essentially ruled out any small bodies existing in the cosmos, stating that:

To make way for the regular and lasting motions of the planets and comets, it's necessary to empty the Heavens of all matter, except per-

Sir Isaac Newton: Brilliant hair and brilliant physics mind.
A terrible meteoriticist.

haps some very thin vapors, steams, or effluvia arising from the atmo-
spheres of the Earth, planets, and comets.

A statement like this from the preeminent physics mind, probably
ever, certainly did not help establish meteoritics as a science of extraterrestrial materials, since it now seemed that small extraterrestrial materials could not exist. As the thinking then went, if there could be no
matter in space other than great bodies like stars and planets, then the
rocks repeatedly reported falling from the sky on peasants are either
(1) from Earth, or (2) made up by said peasants for some short-lived
attention. Even during a time of great scientific revolution, such as the
Scientific Revolution, this was significant baggage to overcome.

Revolutionary and Enlightening Setbacks

The periods historians discuss as the "Scientific Revolution" (beginning in the mid-seventeenth century) shortly followed by the "Enlightenment" were exceptionally exciting times for thinking folk. An incredible list of people made their mark on history during this time, and inconceivable progress was made not only on myriad scientific and social problems, but also, and maybe more important, how these problems were approached. According to philosopher Immanuel Kant, the motto for the entire period was *Sapere aude*, which translates from Latin to "dare to know," or more loosely yet probably more on point, "dare to think for yourself." However, in spite of all the revolutionary thinking that was happening during "the Age of Reason," there really wasn't much revolutionary in it for the study of meteoritics, at least in the short term. While it was avant-garde to reject entrenched dogmas, it was also de rigueur to reject supposedly nonsensical stories, so there was not a lot of patience for simpleton farmers that told tales of flaming rocks falling from the sky. That type of thinking seemed like fantasy, and there was no room for fantasy in respectable scientific circles.

The Stones That Made the Difference

History is littered with stories from all continents of falling rocks from the sky, yet for centuries it was not widely agreed upon how this was happening. As chronicled above, Aristotle's philosophy required falling stones come from Earth, Newton's physics had emptied the heavens of all matter but large bodies and comets, and scientists like Lavoisier developed theories based on chemical and electrical experiments where stones could be created in the atmosphere. Given this, it is perhaps no surprise the source of meteorites was unknown to—or at least forgotten by—academia for millennia, but ultimately the evidence kept coming. Lucky for us, it is hard to ignore the hypersonic chunks of rock and metal that keep pounding our planet. Whereas there are hundreds of meteoritic events described in history prior to our understanding of the source of flying rocks, the section below highlights a few of the events that either captured something about society at the time that they fell or nudged the needle toward understanding the true source of meteorites.

Nogata, Japan (May 861)

In the evening of what was probably May 19, 861, the sky lit up and disturbed the rural area near the present-day city of Nogata on the island of Kyushu, Japan. The next morning, villagers came to find the source of the ruckus was a black stone at the bottom of a newly formed pit in the garden of the Suga Jinja Shinto shrine. This event was not only significant to the inhabitants of the area, but it also happens to be the earliest witnessed fall of a meteorite where pieces are currently preserved.* Since the fall was accompanied by well-documented evidence that priests in the area accepted the rock as coming from the sky, one might think this represented a watershed moment in the history of meteoritics. Unfortunately, however, this early recognized meteorite played no role in the global understanding about the source of meteorites. But in more modern and happy news, the Suga Jinja shrine now brings the meteorite out for public demonstration during the Grand Shrine festival every five years. In October 2011, the festival celebrated the 1150th anniversary of the fall, and the Nogata stone led a series of lavishly decorated carts holding the "shrine treasures." Get your tickets now for the 1,200th celebration in 2061.

Ensisheim, Alsace Region (November 7, 1492)

More than six hundred years after the Nogata fall, and coincidentally the same year Chris Columbus was sailing the ocean blue to "discover" a continent that had been inhabited by humans for over 20,000 years, the Western world was literally rocked by the arrival of the Ensisheim meteorite. The Ensisheim fall occurred shortly before noon, catching people out in the fields of what is now far eastern France to witness its

---

* Subsequent scientific investigation showed that the Nogata meteorite is what is known as an "ordinary chondrite" and was just under 500 grams. More on meteoritic designations can be found in the appendix.

Top: *Painting of the Stone of Ensisheim.* Bottom: *A chip from the famous meteorite. Hopefully it doesn't fly off the page and hit you as you are reading this.*

landing and hear its thunderous boom, which was reported over multiple modern-day countries. The heavy black stone had created an approximately one-meter-deep hole, and after curious townsfolk decided that was no place for their new friend to reside, they promptly removed the rock and immediately started helping themselves to pieces of it as souvenirs. Of its estimated original 135 kg, today the largest stone weighs in at about 56 kg, which, even in the metric system is much less than half of its original weight.

As Nogata is for the East, Ensisheim is the earliest witnessed fall in the West where pieces are preserved. However, whereas the fall of Nogata was essentially ignored by everyone but the local population, the Ensisheim occurrence became an instant sensation.* With the help of numerous dedicated artworks, word of mouth, Twitter, and particularly the recent advent of the Gutenberg printing press, news of the fall raced across Europe. Arguably the most important ears that the news fell upon were those of King Maximilian, who was told by his clever advisors that the stone was a sign of God's favor toward him in his recently begun campaign against the French. The pleased Max helped himself to a couple chunks of the stone and then instructed the main mass be taken to the local church as a testimony to what had happened. And as a demonstration of how little was understood about general physics at the time by the local population, the stone was reportedly chained to the church as to not "wander about at night or, perhaps, return to the heavens in the same fiery manner by which it arrived on Earth."

Across Europe, there was little doubt that the fall at Ensisheim had happened—there were too many witnesses and a massive, weird-looking chunk of rock chained to a church as evidence. However, because Europe was comparatively sparsely populated at the time and population centers of more than a few scores of people were not incredibly abundant, other meteorites in the following decades and centuries were not quite so lucky in the observation department. Most reported falls after Ensisheim had limited witnesses, and almost all

---

* At least as instant and sensational as anything could be in the late 1400s.

of these witnesses were rural farmers or other "uneducated" people. This is of course primarily because most people were rural farmers and "uneducated" at that time. Unfortunately, the sentiment in the ivory towers of academia was that a vast majority of people would say anything to get attention. As such, reports of meteoritic interactions in the following four centuries were either completely ignored or cast away as senseless stories from attention-seeking simpletons with a flair for the dramatic. One particularly flagrant example of this unfortunate attitude was captured following a brilliant fireball with accompanying falling stones in southern France on July 24, 1790. Whereas this event was witnessed by thousands of people and stones that fell were collected in multiple villages over a wide area, the "scientific" response to news of the event was published in nearby Montpellier in a report issued by the editor of the *Journal des Sciences Utiles*, Pierre Bertholon:

> *How sad, is it not, to see a whole municipality attempt to certify the truth of folk tales . . . the philosophical reader will draw his own conclusions regarding this document, which attests to an apparently false fact, a physically impossible phenomenon.*

The same year, further insult to the concept of stones falling from the sky was added by Abbé Andreas Xaver Stütz, the assistant director of the Imperial Natural History Collection in Vienna, when he penned a work with the classically dismissive title, *On some stones allegedly fallen from heaven*. This paper discussed a collection of remarkably similar stories from the last few decades to the one above, although across various parts of Europe. Stütz opined:

> *It was said that the iron fell from heaven. It may have been possible for even the most enlightened minds in Germany to have believed such things in 1751, due to the terrible ignorance then prevailing of natural history and practical physics; but in our time it would be most unpardonable to regard such fairy tales as likely.*

And thank you, Herr Stütz, for that unhelpful mix of strange nationalism and intellectual elitism.

However, as steadfast and flippant as these all-too-common sentiments of the mid to late 1700s from academics may sound, there were still a lot of people willing to search for answers using good scientific practice. One such example is Peter Simon Pallas, who after learning about a ~1,600-pound (~725 kg) mass of stone and iron found in the remote area of Siberia south of the village of Krasnojarsk, dedicated much of 1772 to investigating the object. To reiterate, Pallas heard "1,600-pound blob of iron," and "remote Siberia" and thought "road trip!" But it was a productive road trip, to say the least. Pallas thoroughly described the strange object and its surrounding geology, keenly pointing out that there were no iron-smelting operations anywhere near the region and moving such a large mass of iron to that location would have been essentially impossible, especially given the remoteness and terrain of the area. Additionally, that part of Siberia is more than 1,500 miles from the nearest active volcano, discounting the theory that the iron mass was burped out by a volcano. Oh, and as it so happens, the native Tartars believed the mass to be a holy relic that fell from heaven. Pallas did not openly say that he agreed with the Tartars, but the alternatives were quite limited. After Pallas distributed samples around to the scientific centers of Europe, somehow the focus of the debate was not whether it came from the sky or not, but whether the iron mass was natural or if it had been previously smelted. One daring scientist named Franz Güssman was bold enough to suggest that the huge Krasnojarsk stony iron object was "flung down from above," citing similarities with the Hraschina iron mass that was witnessed to have fallen in 1751 in Croatia. Güssman wrongly pronounced that the samples had been carried into the atmosphere and inflamed by lightning before they fell . . . but at least he was on the right track that they were falling from the sky.

In addition to a number of unexplained oddities apparently falling from the sky around the globe, it was not lost on many of the more open-minded scholars of the time that the 1781 discovery of the planet

Above: *A drawing of the main mass of Hraschina, the iron mass seen to have fallen from the sky in 1751.* Below: *A piece of the Krasnojarsk mass, seen to be absolutely gigantic and in the middle of rugged Siberia, where nobody was making iron. Any ideas on how it could have gotten there? Anyone?*

Uranus,* the first such discovery since antiquity, showed that space was not fully mapped or understood. And as the population covered more and more of the surface of the Earth (particularly Western Europe,

---

* I remember laughing aloud when I read in Bill Bryson's *A Short History of Nearly Everything* that William Herschel wanted to give his new planet the name "George" in honor of King George III. Had that happened, I guess the mnemonic saying to remember planetary ordering would then have to be "My Very Eager Mother Just Sold out in order to gain favor with the King; I can't believe we have a stupid planet named George."

which for good reason thought itself intellectual ground zero), more and more rocks falling from the sky were noticed and cataloged. These stories over the ages had too much in common to be ignored.

This was especially true for the German physicist and musician Ernst Chladni. In early 1794, Chladni published the revolutionary (yet clunkily titled) *On the Origin of the Iron Masses Found by Pallas and Others Similar to It, and on a Few Natural Phenomena Connected Therewith,* which really set the field of meteoritics into motion. The time had finally come for the idea that rocks falling from the sky were coming from outer space. And while this idea was held back by the world's greatest philosopher (Aristotle), the world's greatest physicist (Newton), and the world's greatest chemist (Lavoisier), it would eventually be the thoughts from a master in acoustics that would resonate the loudest and truest on the origin of meteorites.

In his work, Chladni masterfully disputed the common (and spectacularly incorrect) explanations of fireballs, while compiling and discussing the incredible similarities of eighteen witnessed falls spanning multiple centuries, languages, and regions of Europe.* The major theses of his work were:

1. masses of stone and iron can fall from the sky
2. fireballs are created due to frictional deceleration as said stone and iron materials enter the atmosphere
3. these materials originate in outer space and are not products of earthly processes, but were possibly from failed or disrupted planets

Whereas Chladni did make some now-recognized errors in his work, the main ideas have more than withstood the test of time and still rest as the foundations of the field of meteoritics.

As one might expect, Chladni's revolutionary ideas crashed straight

---

* Since news of the remarkable 1790 fall in southern France (the stones from this fall are named Barbotan for a nearby village) discussed above had not reached him, this was not even included in his report.

Ernst Chladni. Master of acoustics, father of meteoritics, and steadfast proponent of the peacoat.

into the face of conventional scholarly wisdom, and the responses and reviews for his work were almost uniformly negative.* But Chladni had time and truth on his side, and in short order the gravity overlords provided multiple important and timely pieces of evidence that convinced many other academics to join the meteorite party.

Siena, Italy (June 16, 1794)

Just a couple of months following the publication of Chladni's seminal work,† the lovely city of Siena, Italy, received an otherworldly gift. At around seven o'clock in the evening, a rapidly approaching cloud from the north produced tremendous rumblings, red lightning, and a large

---

* In addition to his revolutionary contributions to meteoritics, Chladni is widely considered the "father of acoustics." However, even though he is considered a giant in both of these fields, Chladni never even received a university appointment. Until his death in 1827 at the age of seventy-one, Chladni had to earn his living by giving lectures and concerts while traveling over Europe with horse and carriage.

† It is important to note that Chladni's book was only published in two cities: Riga, Latvia, and Leipzig, Germany. Since information did not get around all that fast in the 1700s, you can imagine that it took quite some time for his ideas to really spread around the academic world, making the timing of the Siena fall even more interesting and important.

shower of variously sized stones that fell on the outskirts of town. Crucially, this spectacular event was witnessed by hundreds of people, including local farmers, children, tourists, and university professors alike, meaning it was not going to go down as a made-up tale from a lonely attention hound. Immediately following the event, a well-respected Sienese professor named Ambrogio Soldani began cataloging reports and collecting samples from the fall. Astonishingly, within three months of the fall, Soldani published an almost three-hundred-page book called *On a Shower of Stones That Fell on the 16th of June at Siena*. According to Ursula Marvin, the foremost authority on the history of meteoritics, Soldani's book "decisively raised the topic of fallen stones from the level of folk-tales to that of learned discourse." It is for this elevation of the topic that Marvin argues the fall at Siena was the most important historical meteorite fall.

Incredibly, just eighteen hours prior to the Siena fall, Mount Vesuvius, about two hundred miles (320 km) to the southeast, started erupting. Explosive volcanic eruptions are not that frequent on human time scales, nor are large meteorite falls, and now both were happening in the same general area at the very time we were realizing falling rocks are real, and finally legitimately trying to find out where they come from. Given these semi-paired events, one popular explanation for the now accepted falling rocks from the sky was volcanic activity. And while some working on the problem thought the rocks could have come from the "obvious" choice, Mount Vesuvius, others also raised the possibility that they came from volcanic activity on the Moon, which had been reportedly "witnessed" by William "Planet George" Herschel in work published just a few years earlier. And while both of these volcanic ideas were ultimately incorrect,* at least academia was thinking creatively about ways to explain rocks from the sky and not just dismissing them as imaginary objects or atmospheric concretions. The conversation had finally been started, but one step at a time.

---

* A lunar volcanic source for fallen stones was a popular idea for quite some time, as it allowed for the noticeable lack of a terrestrial volcano at many places where stones were falling.

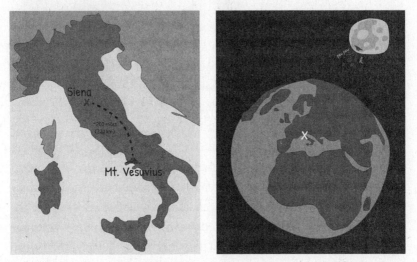

Two possible (yet wrong) volcanical ideas of how rocks came to fall from the sky in Siena, 1794. Left: A "nearby" volcano. Right: Volcanoes on the Moon.

Wold Cottage, England (December 13, 1795)

As we all know from watching decades of James Bond movies, most things of consequence in this world either occur on British soil, or feature someone with a posh British accent. This stands true in the history of meteoritics. If the Siena fall started the intellectual ball rolling, said ball received an important goosing late in 1795 in the form of the meteorite fall at Wold Cottage. On December 13 at around 3:30 P.M., an English plowman named John Shipley was outside in just the right place at just the right time. He was lucky enough to be one of the very few, if not the only person in history, to be sprayed with mud thrown up from a meteorite fall. Shipley was not the only witness to the fall, just the most proximal. Two other farm hands saw the event unfold in all its glory, and several others heard the ruckus the meteorite caused during its Yorkshire-bound plunge. The stone turned out to be a 56-pound (~25 kg), meteorite which still ranks as one of the larger falls in Europe. But what made Wold Cottage a truly important nugget in the history of meteoritics was not the size of the fall, or the location; it was the subsequent literature concerning the event.

Importantly, the owner of the land on which the Wold Cottage meteorite fell was one Captain Edward Topham, an accomplished writer and journalist known for his fashion-forward and mannerly approach to life among the English elite. His well-regarded standing in society was certainly a benefit for him during his life, but it turned out to be especially important for the future field of meteoritics as well. As soon as Topham became aware of the occurrence on his Yorkshire estate, he made haste in doing what good journalists do—he started asking questions and writing about what happened. He conducted interviews from many eyewitnesses, and those who had come to see the rock shortly after it fell. The themes that kept coming up from his interviews:

○ It was a day with no sign of thunder or lightning, yet the sound at the arrival of the stone reminded people of guns firing at sea.
○ Many witnesses noted the strong smell of sulfur emanating from the fallen stone.
○ The stone had a texture "of gray granite" which is not present in the Yorkshire rolling hills in which it fell.

In February 1796, about three months after the event, Topham's detailed description of the event from interviews and sworn testimony of multiple witnesses was published in a local newspaper. Whether or not Topham understood the scientific significance of the event, he certainly understood the significance of media attention to spread the word of important happenings to the masses, and just a small write-up in a local newspaper was not his style. As such, Topham transported the Wold Cottage stone out of the sleepy corner of Yorkshire where it fell to London, where it was prominently displayed (with his supporting literature) opposite a popular coffeehouse in the famed Piccadilly district. For the low, low price of a single shilling,* you could get a view of the

---

* A shilling was a unit of money equal to 48 farthing, or twelve pence, or if you prefer, one-twentieth of a sterling. If that all sounds confusing, it is. In 1796, a shilling would have been just slightly more than the fancy coffee across the street.

otherworldly stone and a handbill of information about how it arrived to Earth.

Concurrent with the fall of Wold Cottage and its associated fanfare, courtesy of the respectable Captain Topham, another man with an incredibly reputable (and British) sounding name, Edward King, was also moving the ball forward on the study of fallen rocks. In addition to his regal name, it certainly did not hurt that Edward King was a fellow of the Royal Society and of the Society of Antiquaries. Following reports of stones falling from the sky in Siena, King started thinking about and compiling various reports of fallen stones over history. In his thirty-two-page work, aptly titled *Remarks concerning stones said to have fallen from the clouds, both in these days, and in ancient times,* King succinctly describes the events of the then recent fall of Siena, but also catalogs the many reports throughout history of falling stones. King delayed publication of his book twice. The first was to discuss the events at Wold Cottage, in which he, like Topham, sets out the geologic differences between the stone and the rocks it fell around:

> . . . *no such stone has ever been found, before this time, in Yorkshire; or in any part of England. Nor can I easily conceive that such a species of stone could be formed, by art, to impose upon the public. Whether, therefore, it might, or might not, possibly be the effect of ashes flung out from Heckla,* *and wafted to England; like those flung out from Vesuvius, and (as I am disposed to believe) wafted to Tuscany, I have nothing to affirm.*
>
> *I wish to be understood to preserve mere records, the full authority for which, deserves to be investigated more and more.*

The second delay in publication was because King received a translation of Chladni's book that suggested that the falling rocks of interest were not of this planet. Eventually King did publish, and while he openly

---

* Heckla, or Hekla, is a periodically active Icelandic volcano. Iceland is home to the closest active volcanoes to England, yet Hekla is still somewhere around 1,000 miles (1,600 km) from Yorkshire.

Above left: *The world's only monument erected at the fall location of a meteorite.* Above right: *The stone it celebrates: the Wold Cottage meteorite of 1795.* Left: *Captain Edward Topham, the owner of the land and man wearing the fabulous slim-cut double-breasted suit, used his media savvy, likely involving that quill pen, to make the fall famous.*

admits that he and Chladni do not fully agree on the source (King preferred the idea that the falling stones were sourced from earthly volcanoes), he noted that Chladni's "facts, which he affirms in support of his ideas, deserve much attention."

Even though King ultimately came to the wrong conclusion, his landmark work was a critical catalog of some of the stories of fallen stones and some crucial observations about their likeness to one another. Importantly, King's work was the first book about meteorites to be written in English, and as such, it was read more extensively by, well, English speakers.

King's book initially received a decidedly negative (and a decidedly elitist and chauvinistic) review in the July 1796 edition of the *Gentleman's Magazine* for King's willingness to admit "the evidence of a few peasants or women." Gulp. Bad review or not, King's work was a success and was widely read. In one particularly stimulating twist, King's book inspired another English naturalist, William Bingley, to report on the fall of a stone he had witnessed in Pettiswood, Ireland,* in 1779. Bingley wrote that he had not previously reported what he had seen because he feared ridicule. Directly referencing Topham's news of the Wold Cottage fall, Bingley wrote into, you guessed it, the *Gentleman's Magazine* with his impressively astute, somewhat erotic thoughts about the heritage of meteoritic materials.

> *I am not without hope, that, upon farther investigation by the learned, my cake and Captain Topham's loaf will be found to have both been baked in the same stupendous oven, according to the due course of Nature.*

Weird sounding or not, Bingley was onto something. His cake was indeed baked with Topham's loaf, Pallas's muffins, Soldani's panettone, and Chladni's dampfnudel. They were all made from extraterrestrial

---

* The Pettiswood meteorite is one of only six recognized meteorites from Ireland.

material, material that is distinct from the rocks we are familiar with here on Earth, and professionals and nonprofessionals alike were starting to take notice.

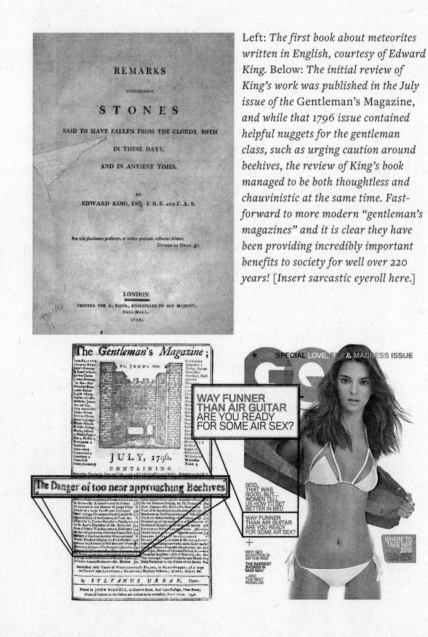

Left: *The first book about meteorites written in English, courtesy of Edward King.* Below: *The initial review of King's work was published in the July issue of the* Gentleman's Magazine, *and while that 1796 issue contained helpful nuggets for the gentleman class, such as urging caution around beehives, the review of King's book managed to be both thoughtless and chauvinistic at the same time. Fast-forward to more modern "gentleman's magazines" and it is clear they have been providing incredibly important benefits to society for well over 220 years!* [Insert sarcastic eyeroll here.]

Better Living Through Chemistry

Alongside the eyebrow-raising discoveries in astronomy of never-before-noticed planets and growing ideas of otherworldly rocks pelting Earth from above, chemists were also busy doing their part in advancing the meteoritic cause. Toward the end of the eighteenth century, the field of chemistry was still in a primitive state when viewed from a modern lens; however, new elements were being discovered and various gases and compounds were being isolated at a dizzying pace. By the close of the eighteenth century, it was possible to recognize differences in chemical compositions of certain rocks, and as curious scientists are wont to investigate curious objects, strange rocks reportedly falling from the sky and giant hunks of metal certainly qualified. Chemist Edward Howard was the first to document chemical differences in rocks that fell from the sky versus your run-of-the-mill, nonflying rocks. Among other things, Howard carefully utilized a newly developed chemical separation technique to show that rocks reported to have fallen from the sky had a lot of the element nickel in them. Rocks that did not fall from the sky contained far less, and in most cases using his method, no detectable nickel. This finding and his report* on the analysis presented powerful arguments that rocks that fall from the sky have similar chemical signatures, and those signatures were distinct from "normal" Earth rocks.

L'Aigle, Normandy (April 26, 1803)

The efforts of people like Pallas, Chladni, King, and Howard—mixed with well-timed natural events—had certainly started a conversation about falling rocks and where they came from. However, opposition to the idea that rocks could fall from outer space lingered. In 1803, a physics professor in Paris by the name of Joseph Izarn published a 422-

---

* Howard's report was titled *Experiments and Observations on certain stony and metalline substances, which at different times are said to have fallen on the Earth; also on various kinds of native iron.* I just love the descriptive, stream-of-consciousness titles 200+ years ago.

page tome with a rambling title* dedicated to summarizing the current academic opinions—of which there were many—on the origin of meteorites. Izarn was a member of the popular "formed in the atmosphere" camp, yet many others thought they must be volcanically produced or tossed up by hurricanes. One small problem with the hurricane hypothesis would have been the general lack of hurricanes in the times and places where stones were falling, but, it was somehow an idea that had a following. There were a select few who subscribed to an extraterrestrial origin for the falling rocks, but this was certainly not the majority opinion, according to Izarn's writings. That majority opinion, however, was no match for the events to come.

On April 26, 1803, a shower of nearly three thousand stones opened up near the town of L'Aigle in Normandy, France. This remarkable event provided a number of things: First, it likely resulted in numerous pairs of soiled underpants, as the descriptions sound terrifying and involve terms like "three enormous detonations in the sky" and "a shower of fire." Second, the event definitively resulted in abundant samples and eyewitnesses, both of which would become extremely important when the investigatory heroics of the incident began.

Whereas the spectacular fall at L'Aigle was a big deal locally, the major national newspapers in Paris (which was less than one hundred miles away from the fall) were silent on the subject. When word of the fall finally did reach the Parisian academic crowd, a renewed vigor for documentation and centralization (courtesy of the French Revolution) caused the minister of interior to hastily dispatch a young scientist named Jean-Baptiste Biot to Normandy for a bit of investigating and report writing.† As you may suspect, the outcome was not a boring, un-

---

* Okay, here it is if you really want to see it: *On stones fallen from the sky—Atmospheric lithology presenting the advance of science on the phenomena of lightning stones, showers of stones, stones fallen from the sky, etc.; with many unedited observations communicated by MM. Pictet, Sage, Darcet, and Vauquelin; with an essay on the theory of the formation of these stones.*

† While the trip was not a long one, only nine days, Biot's time in the field caused him

important government report that was filed away in a desk drawer. The outcome was a thoroughly researched, comprehensive, and decisive report that was easily understandable to anyone. Quite simply, it was a document that changed the course of science.

Biot began his report with an inspirational "scientific call to arms" about figuring out the mystery behind falling rocks. This easily could have been the inspiration for President Whitmore's famous speech in the 1996 movie *Independence Day:*[*]

> *Scientists of all classes, of all countries, have united their efforts upon this great question, guided, not by jealous rivalry, but by the noble love of truth.*

Following this stirring prose, Biot set the stage for his readers with hard evidence, courtesy of the work of Howard and others, noting the chemical similarities previous fallen stones had in common and how these chemical signatures were notably different from common rocks. Biot then included testimonials of a wide variety of witnesses from the L'Aigle area, most of which had no connection to one another at all. The intentional diversity of witnesses included some who were "irrefutable" (because of their status in society) and as he put it in one case, "a very respectable dame that has no interest to impress anybody." He found that regardless of their social status, the witnesses all told very similar stories of what had happened that day in April.[†] With

---

to miss the birth of his only child. This was likely of no consequence to Jean-Antoine Chaptal, the man who dispatched Biot on the trip. After all, Chaptal was responsible for figuring out how to increase the alcohol content of wine, meaning he is personally responsible for tens of millions of births worldwide, so a singular birth probably didn't seem all that important to him.

[*] Just to be clear, in *Independence Day* we were not facing annihilation by meteorites; in that movie we were facing annihilation from aliens.

[†] In his report, Biot compared the chances of such unknowingly coordinated stories to the likelihood that a flash mob of "Thriller" would break out in the town 180 years before the song was written or the dance conceived. At least he should have.

these extensive interviews and some fieldwork, Biot made maps of where stones fell (a near-perfect ellipse) and documented widespread reports of the sounds that accompanied the event. Of course, he also collected samples of weird-looking rocks dissimilar to anything in the local geology. To close his report, Biot reported on the chemical analysis of stones from L'Aigle and showed how they fit the same pattern as that noted in the earlier chemical evidence. Rocks from the sky were just simply different from rocks not from the sky. The Biot report had spoken, and people were starting to listen.

It can be, and is, debated which person, event, or report was the true turning point for the science of meteoritics. Some say the Siena or Wold Cottage falls, some the chemical analysis of Howard and colleagues, some say Chladni's book, and yet others point at the Biot report.* But as Matthieu Gounelle of the Muséum National d'Histoire Naturelle keenly points out in his 2006 work on the subject:

> [*Meteoritics*] *was not born only in a library at Gottingen where Chladni carefully peeped at ancient and modern sources, nor in a chemistry laboratory in London, but also on a road to L'Aigle . . . and evoked by the delicate words of Jean-Baptiste Biot.*

Modern meteoritics plays a huge role in science, as it sits at the crossroads between many already broad disciplines, such as astronomy, chemistry, and geology. However, even after meteorites were recognized as scientific objects originating from outer space, the study of meteorites was still considered only a minor pursuit. It was not until our sights were set for landing on the Moon that investigating meteorites was regarded as a respectable use of time. As such, the real push in the study of meteorites is limited to the last fifty to sixty years of research. In these more recent decades, however, we have learned an

---

* Thanks to Stephen Colbert, I always read this like his old TV show *The Colbert Report*, with a soft "t" on both.

Left: *Jean-Baptiste Biot.* Below: *The meteorite shower over L'Aigle in Normandy, France, 1803. The death knells for meteoritic ignorance.*

almost inconceivable amount about our Solar System from these left-over scraps of rock.

It was quite an adventure for meteorites to be recognized as the treasure trove of information they are, and scientists continue to demonstrate that these space rocks contain answers, or at least clues, to some of the biggest questions we ask about our surroundings and ourselves. But what is it about meteorites that allows us to address such questions? Why do these mysterious rocks from space contain answers to questions we cannot answer from common rocks indigenous to Earth? The abridged answer* is that meteorites are ancient and largely unchanged—certain types have never been melted since they formed over four and a half billion years ago, and thus are excellent time capsules for the genesis of the Solar System. The most primitive types of meteorites are so pristine we essentially sample an unadulterated version of our parent molecular cloud from which the Solar System formed. And it turns out that the seemingly simple ingredients of gas, ice, and dust in that molecular cloud set us on the winding and incredible path we took to rocky planets, water, and life.

---

* For a more detailed answer as to why meteorites are such incredible vessels of information, the curious reader is pointed to the appendix of this book.

Five

Ingredients for Success

To study meteorites is, among other things, to study the history of the Solar System. And from a human perspective, the list of significant events from the Solar System's ~4.5-billion-year existence undoubtedly includes the emergence of life on Earth somewhere near the top. Life emerged on our planet somewhere around 3.8 billion years ago,* meaning that for over 80 percent of the time Earth has existed, it has been hosting some form of life. That is a long time, but by what means did life emerge in the first place? While the "hows" and "whys" of the emergence of life on Earth are questions that have vexed philosophers, religious leaders, and scientists alike for ages, let us touch on the considerably less discussed emergence of *ingredients* as an important prerequisite to the emergence of life on our blue marble.

Ingredients are just raw materials mixed together to form something else. Of course, many people think of making cakes or cookies when they think of ingredients, but making things like plankton, humans, or

---

* Give or take a few hundred million years, depending on whom you ask. Speculation for the first sign of life extends back to about 4.2 billion years ago, but there is hard evidence by 3.5 billion years ago that Earth was inhabited.

cuttlefish is arguably more complex than making a cake. Since chemical elements are ingredients that cannot be broken down into any simpler chemical species, that may seem like a good place to start when assembling stuff to build life. However, the bottleneck to the formation of life is not necessarily the presence of certain elements; the key is how elements like carbon, hydrogen, oxygen, nitrogen, phosphorus, and sulfur* are concentrated and combined with one another. For example, despite having plenty of all those six elements, the Sun has significantly fewer cuttlefish than Earth, demonstrating that the simple existence of specific elements is not the only key for life to develop. Perhaps the most effective way to narrow the scope of important ingredients is to outline the conditions we think are important for life to exist. Of course, one important caveat when we talk about things that we *think* are essential to life is that everything we know about life comes from life on Earth.

There are fascinating thought experiments and research in the field of astrobiology, probing what "other" life in the cosmos could be like—but all life we know of was/is constrained to conditions and environments that have been or are present on Earth, using ingredients found on Earth.† This information is what we tether our ideas about life to, and one can imagine that being a problem as we search for different types of critters in different environments. Life may exist elsewhere under vastly different conditions, conditions we may not currently think are suitable for life. Both the Solar System and neighboring stellar systems are filled with, at least to us, incredibly strange environments. Within the Solar System, things like methane lakes, ice volcanoes, and subsurface oceans exist,‡ and just within our

---

* These six chemical elements are universal to all known living things. They are commonly given the abbreviation CHNOPS (for their chemical abbreviations), and since the first four make up greater than 96 percent of the biomass on Earth, the PS may seem like exactly that.

† For a much more thorough discussion of the formation of life on Earth and how it may exist elsewhere in the cosmos, I direct you to the *Astrobiology Primer 2.0*.

‡ Saturn's moon Titan has methane lakes, Jupiter's moon Europa has a subsurface ocean, and many bodies in the outer Solar System, including the former planet Pluto, have what appear to be ice volcanoes.

near neighborhood of stars there are thought to be bodies that rain iron droplets and planets that have cores made of diamond. As such, it is understandably difficult to predict how/if life might occur in these types of unfamiliar environments.

Yet regardless of the exact types of environments life has emerged under, on Earth or elsewhere, life must always obey the fundamental laws of physics and chemistry. While we absolutely should let our imaginations run wild with ideas about extraterrestrial life, it is important to always keep in mind certain physical barriers. For instance, any reasonable definition of life includes complex molecules, and the overwhelming majority of complex molecules cannot stay bonded together in temperatures above a few hundred degrees Celsius. Simple organic compounds, such as the amino acids that are common to all known life, completely break down as temperatures get above 200–300°C. Many other common molecules found in known life break down at even lower temperatures. Furthermore, our definition of life requires chemical reactions, and "extreme" environments can restrict chemical reactions. No matter how exciting it might be to try to imagine life existing in conditions such as on the surface of a star, close to absolute zero,* or in the almost absolute vacuum of deep space, if no chemical bonding or chemical reactions are taking place, our current definition of "life" is impossible to achieve. But even though these are incredibly stimulating things to think about, we are, for the moment at least, only concerning ourselves with the development of life on Earth. And since we have hard evidence that the environment of Earth has been quite amenable to chemical reactions for its entire recorded history, we are left pondering the provenance of complex molecules on our planet if we want to understand how life developed here.

---

* This is the physically impossible temperature to reach of 0 Kelvin (-273.15 °C, or -459.67 °F), where atoms stop moving. Methinks the mountains on a Coors Light can would be *really* blue at 0 K, but I guess we will never know.

What Are "Complex Molecules"?

A molecule is simply a grouping of atoms bonded together. This can be combinations of atoms of a single element, such as two oxygen (O) atoms that make up the O_2 in our atmosphere or the sixty carbon (C) atoms in the soccer-ball-like C_{60}, better known as buckminsterfullerene, or this grouping can involve different elements like the combination of hydrogen (H) and oxygen that makes up water, or H_2O. At some point, when a bunch of things are strung together, which frequently involves the element carbon, we start calling these things complex molecules. Ignoring the name, complex molecules can actually be fairly simple with only a few elements involved. Alternatively, they can be enormous strings of complication that comprise the entire genetic code of a human being, as in the case of deoxyribonucleic acid (DNA). Simple or complicated, molecules that contain the elements carbon and hydrogen are deemed to be "organic."* Within this category are commonly known macromolecules, very large molecules, like carbohydrates, proteins, and nucleic acids. Because all known life is carbon based, organic molecules turn out to be *the* essential ingredients for life and are almost ubiquitously called the "building blocks" of life in any textbook or journal article on the subject. Thus, our scope narrows again to look for the source of these organic building blocks.

At this point, we can safely state a couple of important things. First, complex organic molecules were required for life to develop on Earth, and second, complex molecules cannot exist at temperatures above a few hundred degrees Celsius. With these knowns, we find ourselves at an interesting place as we try to determine the source of the complex organics on our planet. There is little debate that the Earth, at some point during initial formation, and again after the Moon-forming impact, was entirely molten with surface temperatures well above 1000°C. As such, chemistry and physics dictate that, had they already existed on

---

* This is different from the "organic" agricultural push of the last few decades to avoid artificial chemicals in food. Because many artificially produced chemicals contain carbon, this is a very confusing choice of terms for chemists concerned about what chemicals they are putting in their bodies.

the early Earth, no complex organic molecule would have survived the Moon-forming impact. The Earth flash-melted and became a giant ball of molten rock, a catastrophic process that would have instantaneously evaporated any ocean that might have existed. So, if massive colonies of hyper-intelligent cuttlefish had developed on our planet within 50–100

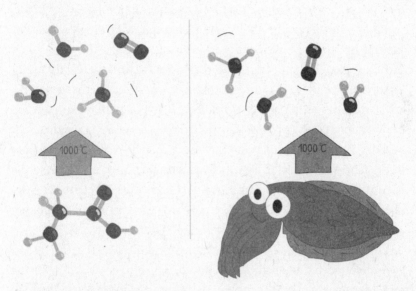

The fates of the amino acid glycine ($C_2H_5NO_2$) and of cuttlefish are very similar when heated to 1000 °C. They both become a gas.

million years after the beginning of the Solar System, they, and all of their hyper-intelligently organized complex organic molecules, would have been disaggregated into their most basic constituent phases in the fiery heat when Theia met the proto-Earth to form the Moon.

And here we are at that oh-so-interesting question: if complex organic molecules had to be available for the first living organisms to self-assemble and for life to emerge on Earth following the Moon-forming impact, where did these organic materials come from? By a relatively simple process of elimination, the origin of these ingredients was either: (1) Earth, (2) not the Earth or, (3) some combination of both. The answer is not only academically interesting but has far-reaching consequences for our understanding of the development of life, and for the probability of finding it elsewhere in the Universe.

Earth as the Source of Complex Organic Molecules

One possible origin of building block organic molecules on our planet was that they were produced on our planet following Moon formation. For this to happen, conditions must have been present that allowed for organic molecules to be synthesized abiotically (without life involved). This is a testable hypothesis, assuming we can re-create in the lab the environmental conditions present on Earth over 4 billion years ago and can wait long enough to see if organic molecules start to populate our beakers. For such an experiment, re-creating any specific environment in the lab is relatively straightforward; however, the far bigger hurdle is knowing what environment(s) to re-create. There is plenty of speculation about what the early Earth was *probably* like, but in reality, we simply don't know. Luckily, scientists are eagerly on the case.

Without outlining all the different organic molecules that can be created under such-and-such conditions, it can be summarized by saying that it *is* possible to make fairly complex organic molecules without using living things to do it. For instance, in one of the first origin-of-life studies ever published, the famous "Miller-Urey experiments" of the 1950s showed that a variety of amino acids, organic compounds, and ammonia can form when electric discharges interact with things like water, nitrogen, methane, and hydrogen gas. Likewise, similar abiotic reactions can occur to produce complex organic molecules under the conditions present at hydrothermal vents on the ocean floor. So, both in the deep ocean and at its surface, certain ingredients for life can be formed under conditions on Earth that *may* have existed 4.4 billion years ago. Very cool. However, it is not so easy to make everything we need for our current biosphere with abiotic chemistry on the early Earth.

One of the obvious issues is that even though we can engineer abiotic conditions that produce some complex organic molecules, we don't

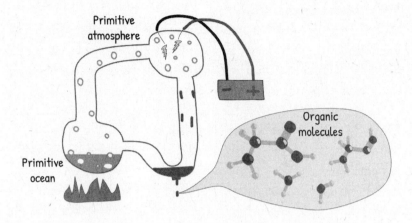

The basics of the Miller-Urey experiment that showed organic molecules can form under conditions that might have been present on the early Earth.

know if those conditions existed in adequate abundance, or even at all, on the early Earth. For example, many of the re-created atmospheres (like the one above) from early experiments that showed organic molecule synthesis was possible contained far more hydrogen gas than we now estimate was available on the primitive Earth. Less hydrogen gas in the atmosphere would have produced far less organic material, with less variety, than was produced in these early experiments that got people so excited about the prospect of abiotic creation of organic material on Earth. Additionally, abiotic reactions are characteristically very slow, and the overall yield of biologically useful products can be very low (and would likely be significantly lower outside controlled laboratory experiments). Lastly, these types of reactions can produce entire classes of similar compounds, with only a select few of them being useful for known life. This issue, known as "selectivity," is like trying to do a 5,000-piece jigsaw puzzle with 10 million pieces from different puzzles all mixed together that basically look the same. And yet you have to complete this one detailed puzzle out of a mass of jumbled pieces, most of which don't fit anything on the puzzle you are trying to complete.

What About Earth's Water?

Having complex organic molecules around to tinker with is obviously important if you plan to build things out of complex organic molecules, but there is another critical ingredient for known life on Earth. Liquid water. And while it is still actively debated whether or not the presence of liquid water is *required* for life to exist anywhere else in the Universe, it is certainly a big deal for Earth's biosphere, and the development of life on our planet. Just for perspective, about half of an average tree is water. Humans are ~60 percent water. And those two examples are things that live on land; percentages of water go up significantly for water-bound species like dugongs or sea cucumbers. So regardless of whether liquid water is required for any life to exist at all, current life on Earth requires loads of H_2O. But where did all of the water on Earth come from? The short answer to that question is that we don't know for certain. The longer answer requires more reading.

Our fellow terrestrial planets Mercury, Venus, and Mars have significantly less water than us, but where most of the water on Earth came from is still an active area of research. The leading hypotheses on the primary source of Earth's oceans are very similar to that of complex organic molecules: (1) Earth, or (2) not the Earth. It is possible that water was incorporated into Earth during the initial accretion process that formed our planet, and we were just lucky enough to keep it around, unlike our neighboring planets. It is also possible that Earth was pretty dry until meteorites or comets delivered water later to the already formed planet, making things like surfing and scuba diving possible in the sloshing H_2O excess. And since certain types of meteorites contain a lot of water, some in upward of 15 percent by weight (some comets have been estimated to be as high as 80 percent water), it does not take that many meteorites or cometary impacts to account for the world's oceans, so this is less far-fetched than it may initially seem.

Similar to any complex organic materials that may or may not have existed prior to the Moon-forming impact, liquid water, if it existed on Earth 4.5 billion years ago, would have had some serious issues when the Earth was flash heated to 1000°C during Moon-formation. Whereas no liquid water, not even if Earth were covered in deep oceans 4.5 billion years ago, would have survived such high temperatures, that does not mean all of it was lost to space. The massive invisible leash that is gravity works regardless of whether Earth is molten or not, and this leash would have allowed Earth to hold on to the majority of its hydrogen and oxygen. The constituent elements of water would have been retained following the planetary impact. And, lucky for us, molecular hydrogen (H_2) and molecular oxygen (O_2) are like old lovers at a high school reunion: as soon as they are near one another, they are going to hook up. Liquid H_2O would re-form without too much fuss as soon as the temperature on Earth was low enough for it to happen. This automatic reassembly of water following a planetary disruption is contrary to the behavior of complex organic molecules. The complex organic molecules required to start life would have had to assemble entirely from the most basic organic ingredients to build life—without blueprints—if it didn't have outside help. Luckily for us, there are clues buried in organic molecules that can help address the possibility of outside help.

The Issue of Handedness

Much like a Lannister always pays their debts, the element carbon always fills its four possible bonding locations. In simple carbon molecules like methane (CH_4), these bonds are basically symmetric: hydrogen bonded around the central carbon in three dimensions looks and acts the same, regardless of its orientation. However, in more complicated molecules, those four bonds that carbon fills can be occupied by four different things. When these cases happen, you can have molecules with exactly the same formula, say the amino acid alanine ($C_3H_7NO_2$), but when viewed in 3-D end up with different orientations of the pieces, and therefore different versions of alanine. These are known as "chiral" objects—objects that look like mirror images of one another but have different properties. This property is generally described as "handedness," as in your right and left hands. They look the same in a mirror,

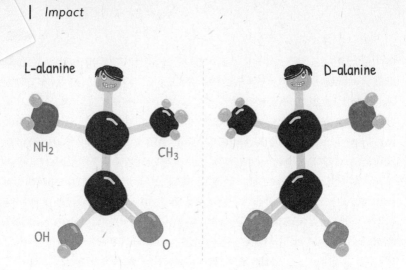

L-alanine

D-alanine

NH₂

CH₃

OH

O

D- and L- molecules are chemical mirror images of one another, but they are not chemically identical, as their constituents are in slightly different relative arrangements. For instance, D-alanine's "left hand" is an NH₂ molecule, whereas L-alanine's "left hand" is a CH₃ molecule.

but one is much better suited for signing your name or piloting a toothbrush.

In organic chemistry, these different-handed molecules are generally designated as the L-type (levorotatory, or left rotation) or D-type (dextrorotatory, or right rotation). The L- and D-types have the same chemical formula, but they are different* and can have different properties. For instance, bacteria use the D-type of the amino acid alanine whereas we use L-alanine. The antibiotic penicillin destroys D-alanine and therefore kills bacteria, and not us, because it only attacks D-alanine. The only chemical difference between L- and D-methamphetamine is that the L-form is the active ingredient in nasal decongestants and the D-form is, well, the unfortunately more commonly known form of meth. A less deadly example is that differently handed molecules can smell

---

* I like to think of this as kind of like tacos and tostadas. The basic ingredients are the same: a tortilla, refried beans, shredded cheese, salsa, and a heap of guacamole. Anything that could go in a taco could also go on a tostada, but they are different dishes that serve different tastes.

different from one another: the difference between the smell of lemons and the smell of oranges is only the difference in the handedness of the oil molecule causing the scent.

The handedness of various organic molecules may have important implications for our understanding of how life developed on the planet. Abiotic synthesis reactions (like the Miller-Urey experiment above) form L- and D-types at the same rate, and a 50/50 mix of the two is the norm for nature.* But it just so happens that life has strong preferences and most life, particularly higher forms of life, is made up of only L-amino acids. Can it really be an accident of nature, a flip of the carbon coin, that life preferentially chose to build from L-amino acids instead of its right-handed cousin?

Space as the Source of Complex Organic Molecules

One of the most basic observations about the origin of life is that, in a geologically short amount of time following the Moon-forming impact, Earth went from zero complex organic molecules to an amount of life significant enough to leave recognizable traces that are preserved in ~4-billion-year-old rocks. We know that *some* of the organic compounds needed to make this leap could have been produced by potential environments on the early Earth, but could a sufficient amount of organic material be produced to start life up? Could this all have happened quickly enough without outside help? Also, did life self-select one handedness over the other, or was handedness preordained based on the building material available? Clues to the answers come from meteorites.

This is analogous to the debate about "spontaneous generation"† that existed until the mid-1800s. The principal idea of spontaneous generation was that living organisms could spontaneously arise from nonliving matter. The classic example was maggots springing to life from

---

* Thanks to thermodynamics, pure L- or pure D- substance will, over time, always become a 50/50 mix of L-/D-types. If the rate of change is known, this reaction can be used to determine how long ago something died.

† This was also a brain-product of Aristotle. Dude was an idea factory.

dead flesh: how else would they get there? No maggot mother was ever seen, yet there the maggots were after a few days. They must have just materialized from the dead flesh, right? Wrong. But it was not until 1862 that Louis Pasteur performed the experiment that definitively answered the question. In his experiment, Pasteur took a nutrient-rich broth and boiled it in a specially designed swan-neck flask that allowed the exchange of air, but not dust (we now know dust contains microbes). If the broth sat unexposed to the outside world, nothing grew in the broth. However, without the swan neck and exposed to the outside world, bacterial colonies started to take over the broth. In this work, Pasteur essentially showed that if something was sterilized and isolated, it would remain sterile.* However, if it is exposed to outside forces, the equation changes and the bugs take over.

This concept can fairly easily be transferred to our entire planet. The immense heat of the Moon-forming impact flash melted and completely sterilized the entire planet by breaking up any complex organic mole-

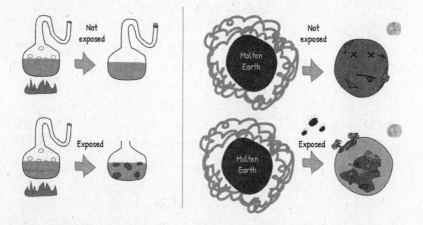

Louis Pasteur's 1862 experiment disproved spontaneous generation of life. Following the Moon-forming impact, Earth was a similar experiment. Luckily, we were exposed to meteorites.

---

* This led to the process called "louization," which is used extensively in food storage and safety. Or is it "pasteurization," I can never remember.

cules that might have been on Earth at that time. So, unless we were exposed to some outside source,* there is no reason to think we would be anything other than a lifeless rock. But the Earth *was* exposed to outside forces following sterilization, and as it turns out, those outside forces can contain a surprising amount of complex organic molecules.

The Amount of Material from Above

Every day in our modern world, an average of more than 100 metric tons of meteoritic material is added to Earth. That is over 70 VW Golfs' worth of meteoritic material added every single day to Earth. That alone may seem like a lot, but that flux was much higher a few billion years ago when there was a lot more material in the Solar System flying around looking for a gravitationally stable home to settle down in. Of course, it is the large pieces that garner the most attention, ending up in museums and on eBay (or causing mass extinctions, of course), but by far the majority of the material added to Earth comes in the form of unassuming micrometeorites and tiny bits known as interplanetary dust particles (thankfully abbreviated IDPs) that are far too small to see with the naked eye.

When it comes to how Earth was built, adding any material to our planet can be significant. However, for this origin-of-life discussion, it is rather important what *type* of material was added, as the ancient sterile Earth needed organic compounds to get the life machine going. And crucially, meteoritic materials, and in particular micrometeorites and IDPs, can contain significant amounts of complex organic compounds, the building blocks of cuttlefish.

A specific flavor of meteorites called carbonaceous chondrites—as the name suggests—contains many carbon-bearing compounds. The relative amounts of carbon-bearing compounds vary significantly be-

---

* As we covered earlier, this "outside source" could also have arguably been an "inside source" where Earth was generating all the organic materials it needed abiotically on its own. However, numerous problems have been identified with this scenario, so an outside source seems required, at least at some level.

tween meteorites, but this is partly due to the amount of heat or weathering (such things destroy many organic components) each particular sample has seen over its >4.5 billion years floating around space and oftentimes bumpy arrival to Earth. On average, carbonaceous chondrites contain approximately 2–4 percent carbon compounds, which is an impressive amount for a space rock. However, some IDPs contain upward of 80 percent organic material! This shows that not only does a huge compositional range exist in extraterrestrial materials, but more important, the early Earth must have received a significant amount of organic material from extraterrestrial sources. Based on the measured amounts of organic compounds contained in various meteoritic materials and their calculated fluxes 4 billion years ago, there were over 275 metric tons (~200 VW Golfs) of *carbon compounds alone* added every day to our young planet! With that amount of complex organic compounds raining down after the Moon-forming impact, it wouldn't have taken too long to build up a respectable inventory of building material to kick off the soon-to-be-developed biosphere.

The Content of Material from Above

While there is little debate that extraterrestrial organics contributed sizably to the carbon budget of the early Earth, when you are trying to initiate life on a fallow planet, the *type* of contributions can be far more important than volume. As an analogy, the most complicated thing I have ever built would probably be the Lego® Death Star™. Had I started with only a mass of molten plastic, I certainly would not have been able to finish the job properly. However, I was lucky enough to have all the premade pieces required to construct the ultimate Imperial weapon, and thus, upon its completion, crush the rebellion of action figures with one swift stroke. If assembly of life followed this Lego® analogy (why wouldn't it?) then meteorites did not just provide atoms of elements that are essential for life (the equivalent of molten plastic); they may have provided many, if not all, of the premade parts to create life.

Of the litany of organic compounds that have been isolated from meteoritic materials, many of them are important building block compounds such as carboxylic acids, amino acids, and ketones. One notable meteorite contains more than eighty distinct amino acids.* Important sugars, such as mannose and glucose, are found in higher abundance in the Murchison meteorite than the element lead is found in the Earth's crust. Of the four nucleotide bases that make up DNA, two have been found in meteorites. Three of the four nucleotide bases that make up RNA have been found in meteorites. Other essential or helpful molecules for life like aldehydes, dipeptides, purines, aromatic hydrocarbons, etc. are found in a variety of extraterrestrial materials. The enormous list of the organic compounds found thus far in meteoritic samples (which grows every year) is a Who's Who of the ingredients for life.

The discovery of a wide array of organic compounds in meteoritic material, of course, leads us to question how such things could exist on rocks that have barreled through space for millions and billions of years.

Present in meteorites

Amino Acids		Nucleotides	Other selected essential molecules for life	
Alanine	Leucine	Adenine (DNA/RNA)	Glucose	Benzene
Aspartic acid	Proline	Guanine (DNA/RNA)	Mannose	Methanol
Glutamic acid	Serine	Uracil (RNA)	Sugar alcohols	Ethanol
Glycine	Threonine		Ketones	Acetone
Isoleucine	Valine		Ammonia	Formaldehyde

A very abbreviated list of some important molecules many nonbiologists have heard of that are, amazingly, present in carbonaceous chondrites and other extraterrestrial materials. This list is ever expanding as techniques improve and new samples are investigated.

---

* Living things use only twenty (or twenty-one) amino acids to perform their daily functions, and many amino acids discovered in meteorites were previously unknown to exist.

How Are There Organics in Extraterrestrial Material?

The four most abundant chemical elements in the galaxy are hydrogen, helium, oxygen, and carbon, in that order. As the fourth most abundant element in the galaxy, it makes sense that there would be a lot of carbon-containing compounds, especially those mixed with other abundant things like oxygen and hydrogen, and other top-10ers like nitrogen and sulfur. In addition to these elements being abundant, all, with the exception of helium,* are chemically very eager to bond. So long as these elements are near one another and it is not too hot, formation of the most basic molecules (one or two elements) takes care of itself. The space between and around stars, or the interstellar medium, has vast areas around –250°C; these areas are prime for the formation of organic molecules. In fact, well over 100 carbon-containing molecules floating around in the interstellar areas have been identified by remote sensing techniques, confirming organic molecules form and exist in these cold environments between stars.

The specific reactions that form each type of complex organic molecule in space is an active and exciting area of research, and not every reaction pathway is fully understood. However, in general, dust and ice grains are primarily composed of very basic molecules like H_2O, CO, CO_2, CH_4, and NH_3. When these basic starting materials are exposed to cosmic rays and/or UV-irradiation, both extremely common things in space, new molecules are formed. When this happens over billions of years to trillions and trillions of ice particles, you end up with a very large number of complex molecules that can be incorporated into meteorites, comets, and other space-faring objects when they form in your stellar neighborhood. In the Solar System, this would have happened (and is currently happening!) somewhere beyond the current orbit of Jupiter, where it is cold enough for ice to persist.† We still have

---

* Remember, helium is a chemical snob, a "noble gas" with a filled outer electron shell that leads to no interest (or real ability) in bonding with any other elements.

† Numerous chemical and isotopic studies performed on carbonaceous chondrites strongly suggest that these rocks formed beyond the orbit of Jupiter. This conclusion is supported by various dynamical reconstructions of the asteroid belt, which is heartening.

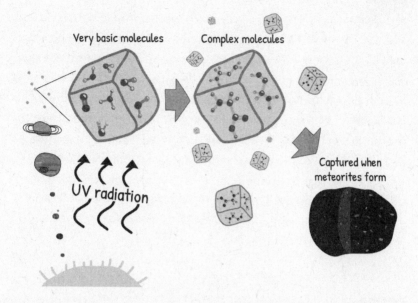

Organics form in the cold parts of space when very basic molecules interact with radiation to form new, more complicated molecules. These can be eventually captured during meteorite formation.

loads of icy particles floating around in the outer Solar System today, and there would have been far more when the Sun was just getting started.

I was astounded when I first learned that meteorites contained even basic organic molecules, and almost in disbelief when I learned major components of our DNA and RNA are embedded in space rocks. Just the existence of these organic molecules in meteoritic material suggests that meteorites played an important role in the development of life on Earth. And even if meteoritic influence was not directly related to the origin of life, delivery of organics to early Earth from meteorites would have, at a minimum, affected the evolution of life and the building materials from which living organisms are made.

Much like a primitive MacGyver, development of life on the early Earth would have used what was available. To that end, the incredible overlap between the list of known organic compounds in meteorites and those found in the biosphere is not likely a simple coincidence. In addition, many of the amino acids found in carbonaceous chondrites contain higher amounts of the left-handed variety, in some cases up to 60 percent more of the L-type than the D-type. The reason why most life developed with left-handed instead of right-handed amino acids could simply be that L-types had an important competitive advantage: there is just simply more of it around. Or, as organics-in-meteorites expert Dr. Sandra Pizzarello and colleagues put it in their 2006 work:

> *Extraterrestrial prebiotic processes might have provided the Earth with a "primed" inventory of essential organic molecules holding an advantage in molecular evolution.*

Additional Bio-Essential Ingredients in Meteorites

As if the simple delivery of the organic building blocks were not enough of a contribution to Earth's early biosphere, meteorites also contain life-critical components that are often in short supply. Once the first critters started living it up on Earth, however that happened, it would have been crucial for them to access nutrients in order to keep the life-party going.

One of the most limiting nutrients to life, even in modern times, is reactive and soluble phosphorus. That is why the fertilizer you dribble on your house plants or farmers spray on their fields is chock-full of the stuff. A typical bag of fertilizer from a home store will be upward of 10 percent phosphorus because it really is what the plants crave. But it is not just plants that crave it; phosphorus is essential to almost all life functions for all creatures. It is an important structural component of both RNA and DNA, and it is the triply important "P" in the molecule responsible for energy transfer in humans, ATP, or adenosine triphosphate.[*]

---

[*] Among many other things, phosphorous is also the literal backbone of vertebrates. The main mineral in bones and teeth is apatite, a highly stable phosphate mineral.

Even though the element phosphorus is incredibly abundant, the vast majority of phosphorus is locked up in insoluble/inaccessible rock forms on Earth, driving the >$15 billion/year phosphate mining industry. Of course, the rub for Earth's earliest inhabitants is that this robust mining industry had not yet been established, making the leap from prebiotic chemistry to living critters potentially even more difficult. As you surely have guessed by now, meteorites may have come to the rescue as a critical source of reactive phosphorus. Because phosphorous in meteorites is in a much more reduced form (formed with far less oxygen), it could—and continues to—easily find its way into biology and the life cycle instead of just being stuck with geology and the rock cycle. As such, early meteorites likely represented an important injection of a much-needed limiting nutrient as life was just getting started.

And lastly, please don't think meteorites of modern times are only good for expensive door stops/collectibles and scientific investigation; they have yet to stop being important delivery vehicles for essential nutrients and raw materials. Even on modern Earth, with the constant recycling of essential nutrients from the massive biosphere and our vigorous mining efforts, there are still nutrient deficiencies in life-forms that are constantly being addressed by meteorites.

Much like phosphorus, iron is an important component of many of life's functions, but it is not easy to get iron in the form it is needed for biology. Even though there are billions and billions of tons of iron laying around basically everywhere, the vast majority of iron in Earth's crust is highly oxidized iron (Fe^{3+}). This type of oxidized iron is essentially useless to biological systems because it is insoluble. Biology needs soluble iron (Fe^{2+}), which is found in vanishingly short supply, particularly in places like the ocean, where organisms get a lot of the nutrients they require from seawater. This lack of dissolved, or bioavailable, iron limits productivity in large portions of the ocean. In certain latitudes, windblown desert dust contains enough iron that can be dissolved when it lands in the ocean to serve the needs of the marine ecosystem. However, large swaths of the ocean, particularly in the Southern Hemisphere, don't benefit much from wind deposition and are reliant on other sources of iron, and you can probably guess this source. Plankton, which are arguably the most important organism on Earth, would be far less abundant and productive if it were not for the delivery of extraterrestrial iron from micrometeorites and IDPs.

The importance of extraterrestrial iron on ocean productivity has caused some researchers to speculate that the well-documented increased influx of extraterrestrial material around 450 million years ago caused a global increase in marine productivity. If such a productivity bump was intense enough, it would have caused a significant drawdown of global CO_2 levels, lowering global temperatures. As such, increased meteoritic delivery may have been the indirect cause of the most intense ice age of the last 500 million years, the Ordovician-Silurian extinction event, which wiped out ~85 percent of marine species at the time. Meteorites giveth, meteorites taketh away.

The Big Picture

It is often said that we are all products of our environments. That statement is routinely applied to situations like when a child grows up in a musical family and learns to play the piano at a high level. They were the product of a musical environment. Sometimes this statement is even applied to biological evolution, like how a certain bird's beak developed to crack a certain type of nut growing in the area. That bird developed, over generations of natural selection, a beak that could access an important food source. But if we are built from organic material delivered from meteorites, then it is also true that we are, ultimately, the products of the environments that existed on ice grains in the outer Solar System 4.5 billion years ago, or on a dust particle floating between stars well before our Sun even formed. And if it was not incredible enough for you that the organic building blocks that our deep ancestors turned into life were delivered from meteorites, consider that the ubiquity of the raw materials, energy required, and the simplicity of creation means that complex organic molecule-rich bodies are almost certainly commonplace in the Universe. If life developed on Earth due to organic materials that formed abiotically in outer space and subsequently were delivered by meteorites to a nurturing environment continually replenished by meteorites, then that suggests that any planetary body in the Universe with reasonable conditions for chemical reactions and a little luck has a reasonable opportunity to develop life, in some form or another.*

---

* This is the basis behind the idea of "molecular panspermia." The basic building blocks should be present for life to develop in many places in the universe, and as such, life should be a relatively common feature around the cosmos simply because there are a lot of places it *could have* developed. Future space missions to potentially habitable (or previously habitable) places like Mars, or various moons of the gas giants like Enceladus, or Titan will nibble at this idea, at the very least.

Six

Free Samples from Mars

The planet Mars is, and has long been, a target of intense scrutiny and intrigue for the human inhabitants of Earth. There are probably a number of factors that explain why Mars over other planets has so excited star-gazers and science fiction writers alike for millennia, but it may largely just boil down to the fact that Mars can be seen regularly with the naked eye—and, maybe more important, it is red. Mars was first observed by the ancient Egyptians ~4,000 years ago, who, noting its red tint, cleverly called the planet *Her Desher,* meaning "the red one." Early Chinese astron-omers termed Mars "the fire star" when they started tracking its motions around the heavens more than three thousand years ago. And of course, the battle-fond Romans gave it the name we use today after their god of war, a direct reference to its bloody coloring. But is this intense interest scientifically warranted, or is it just a psychological trick played on us by the cosmos? Is Mars the planetary equivalent of a mid-1990s cherry red Chrysler LeBaron convertible trying to look like a luxury sports car?

On the Red of Mars

Despite being its visually most recognizable feature, the red of Mars is literally only skin deep. Only the top ~1 millimeter of Mars's soil is actually red, so don't judge an entire planet by its color. Much like the crust of every terrestrial planet in our Solar System, Mars's crust is primarily basaltic rock. But basalt is typically dark gray to black in color, not red, so what gives? In the early days of Mars, its climate was much warmer and wetter than it is now and the interaction of these warmer/wetter conditions with the iron-rich basalt caused the iron at the surface to oxidize.* Since oxidized iron is reddish—oxidized iron is what makes your blood red—this surface oxidation turned a lot of gray iron to red iron. As the planet's climate cooled and dried, the oxidized portions of the basalt became small particles of red dust. These red dust particles have since been blown about by powerful winds on the planet, covering Mars in a very thin veneer of fine red dust. Mars is undeniably red to the human eye, but the most interesting aspects of Mars are contained below its rusty patina.

On the Intrigue of Mars

With thousands of years of unassisted observation, and 150+ years of aided observation, it would be difficult to say when Mars's intrigue was at its highest. In all likelihood, ongoing and upcoming planned missions to Mars will continue to captivate the public with all things Martian, and if/when a crewed mission to Mars happens, even more eyes and minds will be on the Red Planet. While it is difficult to calculate "peak Mars," it suffices to say that intrigue in Mars has existed for a long, long time. The modern craze originated with the Italian astronomer Giovanni Schiaparelli, who, looking at Mars in 1877, believed he saw a system of straight lines on its surface, which he called *canali*.† This observation led

---

* Oxidation occurs when an element reacts with oxygen. In the case of iron, reduced iron (Fe metal and FeO) is silvery to dark gray, and oxidized iron (Fe_2O_3, think rust) is reddish in color. Whereas most of the oxidation likely happened early in Mars's history, its iron is still being oxidized today due to interactions of various minerals when they are exposed to ultraviolet rays.

† In Italian, this refers to "channels," but was translated into English as the much sexier and far more consequential word "canals."

him (and many others) to speculate that the *canali* were engineered by inhabitants of Mars.* Schiaparelli's earliest writings about the advanced civilizations on Mars—they must be advanced if they were building these things—represented some of the earliest documented thoughts about extraterrestrial life on Mars. As additional researchers reported seeing similar things with other telescopes of the time,† the *canali* buzz increased, and by logical extension, the buzz about life on Mars did the same. As a consequence of the *canali* craze, science fiction as a literary genre absolutely exploded, with countless works understandably focusing on Mars or Mars-based creatures. During this gush of new literature, H. G. Wells wrote his incredibly influential *War of the Worlds,*‡ which directly excited multiple generations: first at its turn-of-the-twentieth-century publication to book readers, and even more so when *War of the Worlds* was read as a radio broadcast in 1938 to a panicked American audience, some of whom were unaware it was fictional entertainment and not just an accounting of a very eventful day of news on the airwaves.

However, the early 1900s saw the idea of *canali* on Mars start to fall out of scientific favor. As telescopes and observations improved, the

---

* The great Nikola Tesla was intensely interested in finding ways of communicating with the inhabitants of Mars, and he spent a good deal of time working on this. This work resulted in the first radio-controlled robot, displayed in Madison Square Garden in 1898. This invention set the stage for the radio-controlled satellites and spaceships that would eventually find their way to Mars; so, while Tesla never actively detected signals from Mars or sent signals to an advanced civilization on Mars, he certainly found a way to communicate signals from Mars, albeit more passively than he may have hoped for. To me, this is a great example of how science often works . . . you may not accomplish the goal you are looking to immediately accomplish, but your efforts lead to advancements that, in turn, lead to a general greater understanding. Hooray for science!

† In order to better study the *canali* on Mars by which he was fascinated, the wealthy astronomer Percival Lowell built the Lowell Observatory in Flagstaff, Arizona. The observatory, among other notable discoveries, is credited with first identifying Pluto and remains one of the most important observatories in the world.

‡ As I am always a sucker for a pointed political message, *War of the Worlds* was inspired by the British colonization of Tasmania and was a thought experiment by Wells essentially asking "What if Martians did to Britain what Britain did to the Tasmanians?" Spoiler alert . . . Wells's musings go about as well for the Brits as reality did for the Aboriginal Tasmanians.

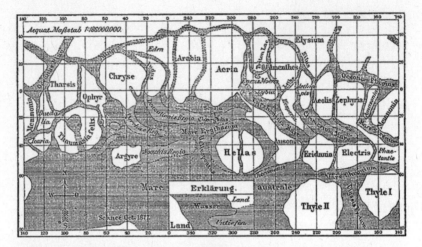

Above: *The original map depicting the* canali *of Mars from Schiaparelli in 1877. Left: As is often said, poor cartography creates chaos, and Schiaparelli's map resulted in multiple fictional invasions of Earth by Martians of various types, the most famous depicted here in H. G. Wells's* War of the Worlds.

canali started to vanish.* It was eventually shown that *canali* were nothing more than illusions from the relatively poor optics of the time, and the *canali* were only as real as the imagined advanced civilizations that

---

* There are channels and valleys on Mars, but they would not have been visible to these early astronomers.

built them. Through the early observation of Mars, the scientific community caused a worldwide stir about the possibility of life on Mars, only to eventually toss a big wet blanket on the idea as better technology emerged. But *canali* or not, curiosity in Mars persisted: from the imaginative writers of science fiction, to the interested public, to countless scientists of various disciplines, intrigue of the Red Planet grew even sharper with the improved telescope optics.

More Modern Studies of Mars

To say that Mars has been a subject of study in science is like saying that cancer is an area of mild interest for medical doctors. After many failed attempts by both the Soviets and Americans, the first successful space mission to Mars was NASA's Mariner 4, which completed a flyby of the planet in the summer of 1965. Since then, more than forty space missions have been attempted dedicated to studying Mars, by space agencies from the USA, USSR (and Russia by itself), the European Space Agency, Japan, China, the UAE, and India. Missions have ranged from trying to understand Marsquakes, to the Martian atmosphere, the amount of water on the planet, and of course to search for extant or extinct life. The amount we have learned about Mars (and the rest of the Solar System) from these missions continues to fill volumes and volumes of scientific journals and books on the subject, and it is certainly not possible to summarize it here. But with scores of billions of dollars spent on scores of missions that have resulted in countless high-resolution images, uberpetabytes of data, and multiple awesome landers roaming around investigating and zapping things, none of these missions have yet returned any physical samples of Mars.

But, hark! We do have physical pieces of Mars! Indeed, we have over 150 kg (330 lbs) of Mars identified and cataloged here on Earth. And it just showed up: these free samples of Mars arrived as meteorites.* At the

---

* They are no longer free . . . you can own a piece of Mars for yourself thanks to the many meteorite dealers that make a living doing this, but the cheapest will cost you at least $1,000/gram.

time of writing, there are >300 approved meteorites of Martian origin, with more being found and recognized every year. The vast majority of these pieces of Mars have been meteorite finds, recovered during meteorite hunting trips in Antarctica or from the deserts of northern Africa and Oman. More important, five of these meteorites (totaling over 44 kg) were actually seen as they fell to Earth.

The first of these falls witnessed by humans occurred in October 1815 in a sleepy area of northeastern France when an ~4 kg stone fell around the town of Chassigny. Another fall occurred in Shergotty (now spelled Sherghati), India, in August 1865, where multiple witnesses recovered ~5 kg of the meteorite shortly after it fell. Northern Egypt joined in on the fun in 1911 when the ~10 kg Nakhla meteorite arrived.[*] About fifty years later, in 1962, the largest witnessed fall yet of a Martian meteorite occurred in Nigeria with the meteorite Zagami. This 18 kg stone implanted itself two feet into the hard ground only a few meters from a likely very startled local farmer. The most recent witnessed sample to fall to Earth from Mars was the 2011 Tissint meteorite. Tissint arrived with multiple loud booms and an accompanying light show in the early morning hours of July 18 in Morocco, with its pieces found starting three months later in the area 48 kilometers from the town of its namesake.

Even though these meteorite falls arrived on Earth in the last two hundred years, they did not come here on a direct path from Mars, and they were certainly not known to be from Mars when they arrived on Earth. It takes multiple scientific studies to determine the source of a meteorite; it cannot be done definitively by hand specimen alone.[†]

---

[*] A person in the area of the fall claimed that it landed on, and vaporized, a neighborhood dog. As sad as this sounds, this story is actually very unlikely to be true: no physical evidence for this was ever found, and language translations of the area and time could easily be interpreted that it landed near a dog and scared it away. Let's go with the second; it is far more likely and it will also keep people from hating meteorites.

[†] Modern-day investigations of Chassigny and Nakhla show that they were launched from Mars about 10–11 million years ago, whereas Shergotty and Zagami spent about ~3 million years cruising the Solar System before they made their crash landings on Earth. Tissint, a 595-million-year-old rock, only floated around space for about 1 million years after it was jettisoned from Mars, taking a more direct path than the others to Earth.

How Do We Know These Rocks Are from Mars?

When collisions in space occur, rocks can break free from their parent masses. If the wayward rock has a high enough velocity to escape the gravity of the main mass (cleverly termed "escape velocity"), it will not return to the surface of its parent rock but will fly aimlessly off on its own until it eventually runs into something else. Earth, in these cases, is that "something else." Most rocks that journey to Earth come from the asteroid belt, where the parent asteroids are relatively small compared to planets, meaning they have much less gravity to overcome and therefore require a lower escape velocity. Mars is not a huge planet—it is only about half the size of Earth—but it is still a planet, and it has significant gravity and therefore a significantly larger escape velocity than any asteroids in the Solar System. It takes a large collision to get rocks off the surface of a planet like Mars, and as such, it was thought to be very unlikely that we would have pieces of Mars come to us in the form of meteorites.*

Yet, these weird rocks in the world's meteorite collections did not fit with anything else. These special rocks had young formation ages, oftentimes billions of years younger than other known meteorites, meaning they were from a geologically active body.† Specific minerals indicated these rocks were in contact with liquid water at some point, another feature not common among "normal" meteorites. These strange rocks had chemical and isotopic signatures very different from other meteorites, suggesting they originated from a very different place. This

---

* The escape velocity to get off of Mars is on the order of 11,000 mph. Volcanic eruptions do not produce rocks flying at these speeds, so the only realistic way this can happen is from large impacts on the surface of the planet.

† There are very few geologically active bodies in the Solar System, and this alone suggests that these samples came from a very large object such as a planet. Importantly, the interplay of (1) planet size and (2) location in the Solar System relative to the Sun are very important when calculating the likelihood of creating potential meteoritic material on Earth. These factors are very much working against having any meteoritic material from Mercury or Venus, but it is dangerous to say something is "impossible." It is, however, unlikely that Earth will ever receive meteoritic material from Mercury or Venus.

troublesome group of meteorites was just simply different from other meteorites. There was speculation in the meteorite community that this group of rocks *could* be sourced from Mars, and this speculation only increased when data from the Viking landers in 1976 showed rocks on the surface of Mars chemically had a lot in common with these strange meteorites. Then, in 1983, scientists showed that the composition of the trapped gases in these unique meteorites was a direct match to the gas in the Martian atmosphere (measured directly during the Viking mission), leaving little doubt that we actually had chunks of Mars on Earth in the form of meteorites.

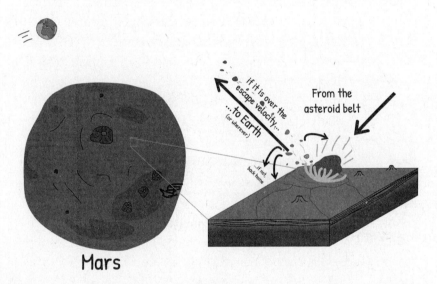

The way rocks make it off the surface of Mars to eventually land on Earth as meteorites from Mars.

After scientists knew which rocks here on Earth were actually interlopers from Mars, interest in these particular samples surged. This cache of rocks allowed scientists to look in incredible detail and learn an incredible amount about the geology and environment of Mars—

science that was and continues to be impossible using remote techniques. Hundreds and hundreds of research papers have been generated on these rocks since; yet, of all the interesting research coming from these pieces of Mars, one study has garnered more attention than possibly all of the others combined.

The Story of Allan Hills 84001

It all started, at least for the human-related part of this story, in 1984 during the yearly meteorite collection trip to Antarctica.* The first meteorite that was found that year, per convention, was named Allan Hills (for the area it was found) and 84001 for the year of the expedition and sample number, leading to its designation as ALH 84001. And ALH 84001 turned out to be quite the doozy of a potato-size rock. It was originally thought that the rock might be related to a class of meteorites called the "diogenites,"† but in 1994 this strange rock was recognized to be a piece of the planet Mars. This was exciting news in itself, as there

---

* More about these incredibly interesting and fruitful meteorite collection trips in Chapter 7.

† This class of meteorites is named after the Greek philosopher Diogenes of Apollonia, who reported the stone that fell at Aegospotami around 465 B.C.E., correctly hypothesizing that falling stones were not from Earth.

were only a handful of meteorites with a confirmed Martian origin at the time. However, the biggest news stories about this sample were still a couple of years away.

On the sixth of August, 1996, NASA released a statement from its administrator, Daniel Goldin. It was a thrilling, yet measured, statement about an upcoming paper in the journal *Science* regarding an exciting potential finding from the meteorite ALH 84001.

> *NASA has made a startling discovery that points to the possibility that a primitive form of microscopic life may have existed on Mars more than three billion years ago. The research is based on a sophisticated examination of an ancient Martian meteorite that landed on Earth some 13,000 years ago. The evidence is exciting, even compelling, but not conclusive. It is a discovery that demands further scientific investigation. NASA is ready to assist the process of rigorous scientific investigation and lively scientific debate that will follow this discovery. I want everyone to understand that we are not talking about "little green men." These are extremely small, single-cell structures that somewhat resemble bacteria on Earth. There is no evidence or suggestion that any higher life form ever existed on Mars.*

—Statement from Daniel S. Goldin, NASA Administrator, August 6, 1996

The following day, President Bill Clinton stood on the White House's South Lawn and discussed the significance of the study.* The president's statement, along with remarks from the authors during NASA's more

---

* Shockingly, the only questions by White House reporters for the president following his remarks on the potential for life existing outside of Earth were, first about (1) the latest Republican-led attempt to restrict abortion, and then (2) his necktie choice for the day. No questions at all about life on Mars. How do you hear that announcement—that we may now have evidence that we are not alone in the universe—and wonder "Where'd ya get that tie?," which is, unfortunately, a direct quote from the presidential press corps that day, and 50 percent of the questions asked. This is not a joke; it actually happened that way. You can watch the exchange on archived video of the press conference.

extensive press conference, were clearly excited, yet also guarded. In particular, statements by Dr. Everett Gibson, second author on the paper, echoed the magnitude of the study but also displayed important scientific restraint, even in the face of a potentially humanity-changing discovery.

For two years, we have applied state-of-the-art technology to perform these analyses, and we believe we have found quite reasonable evidence of past life on Mars. We don't claim that we have conclusively proven it. We are putting this evidence out to the scientific community for other investigators to verify, enhance, attack—disprove if they can—as part of the scientific process. Then, within a year or two, we hope to resolve the question one way or the other.

—Dr. Everett Gibson, coauthor, NASA press conference, August 7, 1996

Dr. Gibson certainly got his wish. Simultaneously with the feeding frenzy that developed in the press after NASA's announcement of potential evidence for life on Mars, scientists from multiple disciplines started to weigh in on the study almost immediately. Many of the arguments were debated, but the most important and most contentious lines of argument from the paper in support of fossil life in the carbonate veins of ALH 84001 are summarized here. Found in ALH 84001 are:

1. Organic molecules called "polycyclic aromatic hydrocarbons" (PAHs), indicative of degraded organic matter from Mars
2. Various mineral grains that are indistinguishable to those formed by some Earth bacteria
3. Bacteria-shaped objects in ALH 84001, and these could be mineralized remains of bacteria that lived on Mars.

Left: *The image that launched a thousand discussions: proposed "microfossils" from Mars in the meteorite ALH 84001.* Below: *Comparison from J. W. Schopf in a 1999 paper of an 850-million-year-old cyanobacteria fossil from Australia* (at top) *with the Mars "microbe"* (at bottom). *They look similar, but the scales are very different.*

Immediately following the release of the manuscript, many scientists started pointing out flaws in these conclusions.* This healthy dose of skepticism compounded over time as hundreds of experiments and studies were launched related to how such "microfossils" could have

---

* This actually happened prior to the release of the article. In addition to the normal peer review process for scientific manuscripts, subject matter experts in areas like paleobiology and evolutionary biology were asked by NASA to evaluate and comment on the landmark research well before, and again during, the August 7 press conference. One such expert was J. William Schopf, who was the "designated skeptic" at the NASA press conference. Schopf has since written extensively on the topic and the unique experience he had. Among other works, his 1999 book, *Cradle of Life*, includes a lengthy discussion of the events surrounding the claims of possible life on Mars from work on ALH 84001, which is worth a read if the intersection of science, politics, and media interests you.

come to be, or how the observations made about ALH 84001 could have had origins other than biologic activity. An extensive list of arguments levied against ALH 84001 hosting evidence of early life on Mars fills entire journal editions, but here are just a few of the clearest examples.

Of the three primary claims of evidence for life noted above, the first two are refuted in a similar way. First, organic compounds such as PAHs *can* be the products of degraded organic matter such as dead trees, bacteria, or iguanodons, but they also form through inorganic processes as well. Such nonbiologically produced organic molecules are common constituents of many other carbon-containing meteorites, such as those discussed in the previous chapter, which are undisputedly produced by nonbiologic activity. As such, finding PAHs is not evidence of past life, it is just evidence that carbon was present in this meteorite and PAHs formed. Similarly, versions of the various mineral grains found in ALH 84001 and used as evidence of past life *can* be produced by bacteria on Earth. However, normal geologic processes routinely form the same mineral assemblages with biology playing no role whatsoever. Again, finding this mineral assemblage is interesting and tells us something about the environment present, but it is—in no way—indicative of life.

Lastly, the images of worm-like objects from ALH 84001 certainly resemble rod-shaped bacteria found on Earth and certainly look like evidence of past life when put side by side with certain types of modern bacteria. This is exactly why these images were the darlings of thousands of news stories about the evidence for past life on Mars. But scale matters. A lot.

The sizes of the ALH 84001 "microfossils" found under the microscope are about 1,000 times smaller than the majority of bacteria on Earth, and about one-tenth of the size of the smallest known bacteria. This may be a bit troubling, but the fact that they are so much smaller than known life is not the biggest issue. The real problem is that there is a physical limit to how small life can be, if it is at all similar to life on Earth, and well, that was the comparison after all. If these ALH 84001 objects were at all similar to bacteria (or really any living cell on Earth), they would need a cell membrane. Such membranes are constructed with a protein-lipid bilayer-protein sandwich, which takes physical

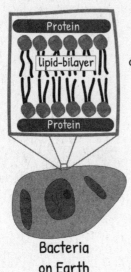 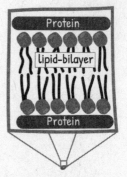

cell membranes are
~10 nanometers
thick

much smaller =
no room for anything else

Bacteria
on Earth

Proposed bacteria
on Mars

It is fun to think of non-Earth life having wildly different forms than Earth-based life, but if it is at all similar, it will have similar physical restrictions. Size is one of them due to the thickness of cell membranes.

space. And the smaller the cell, the more significant amount of space this membrane takes from the inside of the cell. Given the dimensions of the ALH 84001 specimen, the inside of the Martian cell—where all the functions of life would take place—would be ~2,000 times smaller than the space available inside the smallest known living organism.

These tiny sizes may be a bit difficult to picture, because well, they are so small we have no chance of seeing them without a very powerful microscope. But if you think about it on more of a human scale, think of automobiles and the popular Micro Machines replica cars popular in the 1980s and '90s. Micro Machines were ~1:100 scale of cars we might see driving around on the road. In this analogy, this means if a car a human could drive is the normal bacterial cell, the Mars equivalent would be 20 times smaller than the Micro Machine replica, but still somehow able to fit a human driver. Not bloody likely.

As the evidence, studies, and informed opinions rolled in from a variety of disciplines in years following the 1996 paper, it became more and more apparent that the "microfossils" found in ALH 84001 did not meet the community's standards for proof of previous life on Mars. The proposed evidence of life in ALH 84001 was formed by inorganic processes and not from ancient Martian critters. Again, it seems, scientists created, then rapidly quashed the idea of life on Mars. But if you think

If you decide to drop ~$400,000 to buy a Lamborghini Countach, be sure to check that it is the human-size model and not the Micro Machines model. Scale matters.

version 2.0 of the scientific wet blanket on Martian life stifled the public's desire for information about Mars, you are very wrong. If anything, it greatly intensified it.

The Consequences of ALH 84001

Following the ALH 84001 study and the ensuing controversy, enthusiasm for Mars research grew exponentially, creating an even more intense focus on Mars from the planetary science and space exploration communities. Mars missions already under way received increased attention, and multiple missions to Mars were green-lit from a variety of space agencies; funding was redirected to Mars research (much to the dismay of other programs), and the field of astrobiology took off like a Saturn rocket. For better or worse, the 1996 study caused irrefutable changes to the arc of planetary science research.

While it is impossible to list things we might have learned if resources hadn't been redirected to focus on Mars-related science, there are certainly a lot of things that we did learn because of it. For one, in trying to develop a better understanding of what life on Mars *could* look like given its non-Earth-like environs (past and present), scientists sought out and intensely studied extremophiles,* or life at the margins on Earth. This search for "alien life on Earth"—or life that looks and behaves significantly different from life in middle-of-the-road environments—has led to a far greater understanding of where life is possible and what is happening with life in conditions humans deem "extreme." Due to this work that was largely performed after 1996, we now know life exists in places previously thought impossible. Life has been found in extreme acids or bases (ranging in pH from 0 to 12.5), temperatures ranging from -20 to $122°C$ (-4 to $252°F$), and at the crushingly high pressures found deep within the ocean or Earth's

---

* This term is, of course, a very anthropocentric view of what "extremes" are. It may be extreme to us, but for these organisms, their environment is totally normal or even necessary.

crust. Life exists in toxic waste dumps and in hypersaline solutions like the Dead Sea. Certain organisms are even adapted to exist when exposed to very high levels of ionizing radiation, requiring extreme protein protection and cell maintenance to avoid lethal damage or mutations. The amount we have learned about what life can put up with in the last few decades is astonishing. Not only has this multitude of discoveries pushed the "boundary conditions" of life further and further, but it has made the idea of life existing under the previously thought inhospitable conditions on other planetary bodies far more conceivable. And this discussion doesn't even touch on how much this research on extremophiles has increased our fundamental understanding of basic biological processes, species evolution, or the potential (and realized) medical and biotechnology advancements.

And while a better understanding of extreme environments that might be possible to host life is valuable in its own right, observational and robotic missions have actually found evidence that not-so-extreme conditions were present on Mars in the past. Evidence of lakes and rivers of liquid water on Mars has been found, and the clay minerals

The Curiosity rover taking a selfie as it cruises around Mars finding out stuff about the Red Planet, including evidence for its history of habitability.

deposited in that water suggest this ancient environment *could* have supported microbial life. The (in)famous 1996 paper on ALH 84001 did not find ancient life on Mars, but in the long run, it indirectly found out that Mars was a habitable place for about a billion years in its early history.

The Importance of Water on Mars

The most basic things that life requires to exist are: (1) a source of energy, (2) basic materials for building things, and (3) a liquid solvent to move those things around in. Most places in the known Universe do not seem to struggle with the first two requirements, so it is access to the third that is likely the biggest bottleneck for life. And since all life that we know of uses water as that liquid solvent, is it commonly said that life cannot exist without liquid water. Entirely true or not, it certainly seems like that could be the way it is. As such, evidence of the presence of liquid water on Mars is a really big deal for people who care about things like . . . well, life on Mars, or, on a much bigger scale, contemplating our place in the cosmos. Whereas remote techniques have been crucial for qualitatively showing that water once existed on Mars, it is only through the study of Martian meteorites that we can quantify the amount of water that was present, and how much has been lost over time.

Just as scientists can pair information from meteorites with the remote techniques studying modern Mars, we can do the same in investigating the climatic and atmospheric evolution of Mars. Thanks to the Martian meteorites in our collections, we have time capsules of Mars spanning almost the complete history of the planet. Briefly, the geologic history of Mars is broken into three main periods: the Noachian, Hesperian, and Amazonian, each named after areas on the surface of Mars that typify the geologic periods. The Noachian, the era just following formation, was warmer and wetter than it is currently, probably due to a high amount of volcanic activity releasing greenhouse gases and keeping the planet's surface temperature up. During this period, Mars was able to sustain liquid water in its large basins, especially at low latitudes. If life

ever developed on Mars, this first ~500 million years woul
favorable window, by far,* and the type of environment the
verance is exploring. Next, the Hesperian, spanning from ~
3.0 billion years ago, saw the climate turn much dryer and r
The liquid water that was likely once abundant on the surface became
more localized and eventually disappeared through freezing, sublima-
tion, and loss off the planet. Lastly, the Amazonian, the period Mars has
been stuck in for ~3.0 billion years, is characterized by very cold and
hyperarid conditions. Not a great rut to be in as a planet, but it certainly
is still better off than Venus or Mercury.

The Ages of Things

Knowing the age of a sample is one of the fundamental pieces of information needed to reconstruct a geologic environment, whether you are
working on Earth or working on other planetary bodies. That is basically
what geologists are always trying to do: look back in time to figure out
what happened, and then go have a beer (or four) with their colleagues.

The chronologic studies of Mars come in two main flavors: those
done using remote techniques to find out how old various parts of the
surface of Mars are, and those done on Martian meteorites in the lab to
find out how old each individual rock is. The first is done using satellite
imagery, meticulously looking at the amount and size of craters on the
surface and converting that into an age of how long that surface has
been exposed to the wilds of space. The basic premise is that old surfaces have seen a lot of impacts and young surfaces appear fresher since
they have not been assaulted by asteroids for quite as long. Think of it
as a car hood from Missouri, my home state and land of periodic hailstorms. A brand-new car in Missouri will have a nice, undented hood,
whereas a twenty-year-old car will have seen multiple major and a lot of

---

* The meteorite ALH 84001 is from this earliest period, which added intrigue, since it
was thought at the time that this would have been the period to develop life on Mars if
it ever did happen.

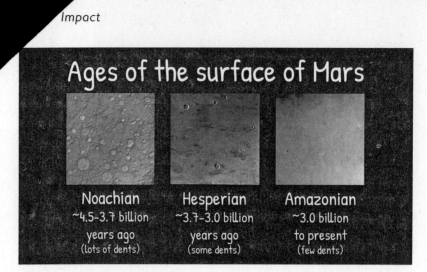

How ages are calculated on the surface of Mars, when no rocks are available from the area. Beat to hell = old, so-fresh-and-so-smooth-smooth = young.

minor storms with plenty of damage to show for it. The same is basically true for the surfaces of Mars: older surfaces are pocked with numerous craters of multiple sizes, whereas young areas of Mars are flatter and only have smaller craters.[*] This method is a great way to determine relative ages of large areas, but a lot of assumptions go into it; most important, one has to know the rate of impacts (or how often "hailstorms" occur), which can only be calibrated if you independently know the age of the rock layer.[†]

The second flavor of chronology, dating meteorites from Mars in the laboratory, is near and dear to my heart from my previous research on this subject. To be honest, a mix of near-and-dear and a little traumatic, since it is so damn difficult and time consuming. Without getting into

---

[*] As discussed in Chapter 1 for Earth, large impacts on Mars are far less frequent than small impacts.

[†] This method for Mars is calibrated using rocks from the lunar surface collected during the Apollo missions and crater counting on the Moon. Scientists have taken the rate of rocks hitting the Moon and mathematically transferred this to Mars, using a fudge-factor for their different locations in the Solar System. This is not a method without uncertainty, but it works well for relative ages.

the gruesome details, the method involves separating minerals from one another in the rock, then measuring the elemental and isotopic ratios of the mineral separates. Because most rocks contain numerous radioactive isotopes that can be used as clocks, this allows the age of the rock to be calculated, alongside other ancillary information you pick up in the process. In most cases, a gram or so of an individual rock is sufficient, but because individual mineral phases have to be separated and cleaned to do this, it can take more than six months of work in the laboratory to determine such an age.* Luckily, there are a lot of good and patient people that do this in the community, and the age range of the >300 Martian meteorites spans from ~4.1 billion years old to a scant 160 million years ago; in other words, for nearly the entirety of Mars's history.

Other than just the ages, this chronologic work on Martian meteorites has provided us with an incredible amount of information about the interior of Mars that we would not possess otherwise. For instance, we know that the multitude of volcanoes on Mars tap different reservoirs in the Martian mantle and that mantle has many regions that are vastly different from one another. Because plate tectonics does not occur on Mars, these mantle reservoirs don't get mixed up like they do on Earth. Also, because the Martian meteorites in our collection span almost the entire history of the planet, we know that Mars has been geologically active† for 4 billion years or more. This fact alone allows scientists to estimate the heat flux of the interior of the planet, informing us about how much radioactivity is present from elements like uranium, thorium, and other long-lived radioactive isotopes, which are important heat sources for planets. Additionally, the chronology of Martian meteorites not only

---

* I can speak from experience here. It took me over three months of staring down a microscope using a tiny pair of tweezers (and many audio books) just to isolate and separate the mineral phases needed to get a crystallization age for the Martian meteorite Tissint. This was the first (and last) Martian meteorite that I did chronology on.

† This means that the surface of the planet is actively being altered, from things like volcanic activity or tectonics. Impacts don't count, because the planet is not really doing anything, it is just passively taking a punch.

taught us that Mars formed as a planet very early, within the first 10 million years of the Solar System, but we also know that it formed a core, crust, and atmosphere somewhere between 20 and 40 million years after the start of the Solar System. Numbers like these may just seem like dates on a timeline, but they are critical for a broad understanding of how planets form and evolve, not only in our Solar System, but also in other stellar systems.

On the Importance of Samples

One of the reasons meteorites from Mars (and other meteorites in general) have been so important for our understanding of the entire Solar System is that Mars is the only other external planet we have samples from. Because Earth underwent a massive Moon-forming impact that melted everything within the first couple hundred million years, erasing a lot of information of its earliest history, it is very difficult to find anything here older than 4 billion years, making Earth a less than ideal place to look for clues about planet formation and the early Solar System.[*] Since the Moon-forming impact, it hasn't gotten much easier, as plate tectonics has repeatedly melted and shuttled most of Earth's rocks around. If we only had rocks from our home planet to look at, we would be far less informed about terrestrial planet formation than we are now.

Over the years, Mars exploration has been focused on remote techniques. And, that is understandable, given the technological and resource challenges of crewed/sample return missions to Mars. We have certainly learned an amazing amount with the plethora of fancy instrumentation (telescopes, orbiters, landers, rovers, etc. . . .), but there are many things that need "ground truth," which can only come from having the sample in a laboratory. There are a lot of things that remote techniques simply cannot do, no matter how cool they look or how

---

[*] The oldest dated sample on Earth is around 4.4 billion years old from the Jack Hills region of Australia, and it is not even a real rock, it is just a tiny (yet robust) mineral leftover from a previous generation of rocks squished in with other rocks.

much they overperform.* The stash of Martian meteorites we have in our collection, and the new ones we will get, are a huge boon to the planetary science community. And while the meteorites we have allow us to really dig into the details of the planet, the downside is we don't really know *where* these meteorites came from on Mars,† and that is a major problem. There is no geologic context, just a chunk from the planet from some unknown random place.

With the recent space exploration success by private companies such as SpaceX, and the continued interest and accelerated collaboration between governmental space agencies, the possibility of returning samples from Mars is becoming ever more tangible. But if/when we get to a point where we bring samples of Mars back home, they will be chosen types of rocks, not just the random rocks from random locations that are launched off the surface and happen to survive the unlikely journey to Earth. The chosen pieces will be carefully selected to get the most scientific value possible from each sample. Given the cultural curiosity about life on other planets, the chosen rocks will likely be parts of ancient lake beds or old riverbanks to better understand past environmental conditions, and of course to look for evidence of past life. And if the meteorite ALH 84001 taught us anything, it is that even with samples in the laboratory, signs of life can be ambiguous, and it takes a collection of scientific disciplines to come to robust conclusions. Without samples, something as important as evidence of life on another planet will be hotly debated by scientists and nonscientists alike until we can get the physical evidence into the laboratory.

The study of Mars over the last 150+ years has been littered with what

---

* Famously, the Mars rovers *Spirit* and *Opportunity* were planned missions that were expected to last just under ninety days. They both landed in January 2004 . . . *Spirit* was active until March 2010, *Opportunity* made it until June 2018. Not bad.

† With the geochemical data from both meteorites and remote sensing, paired with the high-resolution maps of Mars, it is tantalizing to try to locate the specific craters on Mars responsible for these special deliveries to Earth. Thus far, this has been attempted, but not conclusively done. However, the possibilities have been greatly narrowed, and the search continues . . .

some might call "improper conclusions" or even "mistakes." There are no Martian-built *canali* on Mars. But without the poor optics of early telescopes that led to the ideas of *canali*, Nikola Tesla would not have devoted his time to contacting Martians, leading to his development of radio controls. Percival Lowell would not have sunk a fortune into building the Lowell observatory. Creative minds would not have written about life on other planets, inspiring countless dreamers to think beyond Earth. Similarly, we do not yet have evidence that life exists—or has existed—anywhere outside of Earth. But had the "microfossils" of ALH 84001 not stirred up the scientific community at large, how much less would we understand about life on Earth or its possibilities elsewhere? Would we have developed the capabilities to land large spacecraft on other planets? Would interdisciplinary scientific research be at the impressive stage it is today? It is impossible to replay history with these changes, but these "missteps" also provide good examples of what the field of *science* is: an ongoing search to understand the natural world. Science is a process that experiences hiccups: individual hypotheses and individual studies can be wrong, but as long as we continue to search as a scientific community, the field evolves toward a better understanding of our surroundings. Oftentimes in extremely exciting ways.

Seven

From Space to the Laboratory

The results from meteoritic studies can have a profound effect on science and society alike, but how do these samples end up in laboratories as objects of study? Would you be ready on a moment's notice to jump on a plane as soon as you heard news of a meteorite fall? Would the scientific (and monetary) benefit of having the freshest possible sample take you to places where you may end up being chased by corrupt authorities, imprisoned without so much as a phone call, or robbed at gunpoint for a couple grams of space rock? Modern meteorite collectors flirt with these situations routinely when searches occur in unstable parts of the world, which turn out to be a rather sizable portion of the planet when you look at it from the perspective of area.

Much like playing Frisbee with your friends is difficult without a Frisbee, studying meteorites does not happen without meteorites. Of course, there is some fine and important work in the field that is done without samples, but most meteoriticists are very much dependent on actual samples of space rocks. Exactly how (and which) samples are obtained depends on the individual researcher, but from a community standpoint, we need samples, and the more the better. From the discipline's beginnings, if you knew the right people and were interested in the science of meteorites, it was at least possible to get your hands on some celestial material. That has not drastically changed in the last 200+ years; samples are generally available to anyone interested in studying them. But the procurement process of meteoritic material has changed a great deal from the days of just needing a buddy at the natural history museum. The search for more meteorites has prompted multimillion-dollar investments from several national governments looking to expand the community's access to rare samples. Modern meteorite collecting is not just a hobby for rock hounds with a desire for the exotic, but is a full-time, and often very thrilling vocation for many.

For most of human history, and the vast majority of time meteorites have been a subject of study, meteorite collection has been not much more than a game of luck. If you were fortunate, you would stumble across a weird-looking rock that was a meteorite during a hike, or maybe you would turn up something strange while plowing your field.* If you were *really* fortunate, you would see a meteorite fall and then go pick it up—or better yet, almost be hit by one, providing you with both a new souvenir and a fresh outlook on the fragility of life. But these type of witnessed falls, close encounters or not, are quite rare.†

These passive, random encounters did a respectable job over the years of providing researchers, museums, and curio shops around the world with a reasonable stream of meteorites from a variety of types. Before 1955, there were just over 2,000 confirmed meteorites in the world's collections, with ~35 percent of those being witnessed falls. But in the mid-1950s, the world was changing fast and setting its sights on

---

* This has happened a number of times. The discovery of the meteorite D'Orbigny, one of the oldest and arguably one of most important basaltic meteorites, was made in 1979 when a farmer near Buenos Aries was plowing his corn field and hit the 16.5 kg stone. Today, pieces of D'Obrigny sell for ~$500/g.

† To provide some context, let us revisit the first witnessed meteorite fall that we have a physical sample of, the Nogata meteorite that fell in Japan on May 19, 861. In the ~1160 years since, there have been 1,206 confirmed falls (as of early 2021), which if you are good at math, works out to an average of just over 1 witnessed meteorite fall per year, over the entire globe. Of course, this value is significantly front-loaded as the world's population has spread over time, causing more falls to be witnessed.

space. This collective upward gaze to the stars prompted a significant change in how meteorites were viewed. Demand for meteoritic material, both from the public and the scientific community, increased exponentially. While this interest surge had multiple catalysts, the most important was the NASA and the Soviet space programs starting to explore beyond Mother Earth. One corollary to these robust programs was that the equipment and methods, largely being designed to measure the imminent return of lunar samples, were similarly well suited for addressing interesting questions meteorites had some of the answers to. In addition to the ramping up of space programs, another important event in the burgeoning world of meteoritics was the fortuitous falls of the Allende and Murchison meteorites in 1969, which, in perfectly coinciding with the space race, solidified meteorites as prized items to study and collect going forward. And when something turns out to have value to humans, society seems to find a way to obtain more of this sought-after item. Meteorites are certainly no exception.

The Hard Truths of Meteorite Prospecting

One of the important basic facts about meteorites is that they fall all around the globe. Where they fall is unconnected to latitude, longitude, climatic zone, or political boundary; it is effectively just random, all around the world. As such, if someone is trying to find meteorites, for either monetary or scientific purposes, they have to deal with this hard reality of randomness. Another ding is that more than 70 percent of Earth is covered by water, so an equal percentage of meteorites that fall are quickly rendered useless and lost forever. This of course narrows the areas of the Earth to search for meteorites to only the "land" parts, but some additional fine-tuning is still needed to make meteorite prospecting a worthwhile endeavor.

Precious geologic commodities such as gold, diamonds, and oil are naturally concentrated in particular geologic environments. If one understands how those commodities are concentrated and where, one can elevate themselves to positions of "mogul," "magnate," or even "tycoon." And since it does not take an economics wizard to know that it is more profitable to mine one ounce of gold from one ton of rock than it is to mine one ounce of gold from 10,000 tons of rock, finding areas of higher concentration of gold will get you looking and feeling tycoonish far faster. Areas of higher efficiency like this are precisely why economic geologists are trained to search for natural concentrations in the Earth of certain things, and that is why some areas are exploited for raw materials and some areas are not. But meteorites are not your typical geologic commodity, and, importantly, it is not necessarily the general *lack* of meteorites that is the biggest impediment to finding them. The bigger reason why our collections are not overflowing with samples from outer space is the general difficulty in locating and discerning meteorites camouflaged among regular terrestrial rocks.

Complicated geochemical modeling, extensive seismic profiles, and thorough groundwater sampling are principal methods used to locate "normal" geologic treasures, but these are uniformly useless as aids in meteorite location. However, there are some notable (and simple) distinctions between meteorites and normal Earth rocks that can be helpful. Most meteorites are: (1) magnetic and (2) dark in color. While these two generalizations ignore the properties of meteorites from larger bodies like Mars and the Moon, these unusual types of meteorites are extremely difficult to tell apart from common terrestrial rocks, and, as such, require special circumstances and equipment to recognize. On the other side of the coin, obviously not everything that is magnetic and dark is a meteorite, but these characteristics work as a very effective first filter to protect against finding dreaded meteor-wrongs.[*]

Meteorites are more magnetic and are darker than most Earth rocks for a few very simple reasons. Firstly, meteorites in general have more iron than Earth rocks. Iron is magnetic, and since the bulk of Earth's iron is in its core, meteorites have more of it than most surface rocks on Earth, and are thus more magnetic. The reason most meteorites tend to be so dark is slightly more complicated. Not all meteoritic material

---

[*] I would be remiss if this community joke was not made at least once in this book.

is darker than Earth rocks—some meteorites are actually very light in color—but meteorite surfaces are very often dark because of the perilous journey the meteorite experienced on the way to Earth. When meteoroids enter Earth's atmosphere, they are cruising along at incredible speeds. Earth's relatively thick atmosphere quickly slows down the intruder, causing some serious heat to build up around the rock, melting its outermost layer. Throughout its journey through the upper atmosphere, this process continues to ablate liquid rock off the meteoroid until the meteor slows down enough to stop melting itself, eventually landing on Earth's surface. The end result is a smaller rock than when it started into the atmosphere and a 1- to 2-millimeter dark, scar-like crust on the meteorite known as a "fusion crust"—a great tool for determining what is a meteorite and what is not.

Since there are at least a couple of simple, consistent characteristics that set meteorites apart from terrestrial rocks, it is possible to use

The Marília meteorite that fell in Brazil in 1971 is one of many prime examples of a meteorite with a dark fusion crust, and a light-colored interior. The light color of the meteorite is shown at the top of the picture, but this color is largely obscured by the outer 1–2 mm of fusion crust.

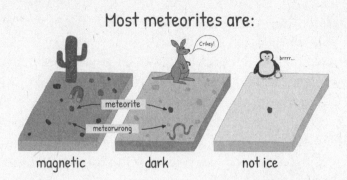

A *cartoon showing the most basic (yet generally effective) ways to tell a meteorite from a terrestrial rock. The easiest places to search for meteorites are places that don't have a lot of stuff in the way to obscure your view. Deserts, particularly deserts with naturally light-colored rocks, and ice fields are good areas to recognize meteorites.*

these as aids in searching for extraterrestrial samples that happen to be lying around among Earthly rocks.

For instance, if you happen to be strolling around the desert and want to look for meteorites during your hike, just walk around with a magnet on a stick. You will likely see many dark rocks, so this "magnet-on-a-stick" method is the easiest way to eliminate nonmagnetic things that are probably not meteorites without bending over hundreds of times. In fact, this is such a famous method, you may even be familiar with the old meteorite prospector's saying: "If your stick doesn't stick, you must not a-quit . . . looking."

Whereas the magnetic method is probably the best choice when there are a lot of dark rocks around, there are vast areas of the planet that have a noticeable lack of dark objects, making the "hey-that-is-a-dark-rock" method a surprisingly useful meteorite location tool. In fact, the most productive meteorite hunting grounds on Earth are areas that (1) naturally lack dark rocks and (2) are devoid of vision-obscuring vegetation. This magic combination makes it far easier to locate meteorites that have fallen over the years. In case you are looking for an adventurous vacation that may help advance science in the process, here is a compiled list of the most important meteorite hunting grounds below.

Nullarbor Plains, Australia

One of the early recognized and particularly fruitful hunting grounds for meteorites is the Nullarbor region of southern Australia. What the Nullarbor lacks in trees, topography, and color, it makes up for in meteorite productivity.

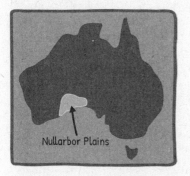

Nullarbor Plains

The flat, treeless expanse of the Nullarbor is the largest single exposure of limestone in the world, so black rocks stick out quite dramatically. Also, because the Nullarbor has an arid climate, and has for tens of thousands of years, meteorites do not degrade like they do in more temperate areas, meaning they tend to hang around waiting for intrepid hunters to spot them. For the true road trip fans, the Nullarbor boasts one of the longest stretches of road in the world without *any* turns. Driving the Eyre highway, you can motor along for more than 90 miles (145 km) without turning the wheel at all. While this may sound like an irresistible trip of a lifetime, it is important to keep in mind that Australian legislation designates all meteorites found in Australia as "property of the Crown"* and all finds must be turned over to authorities; meaning you will likely end up donating your newly located meteorite to one of Australia's fine public institutions. As a consequence of this law, meteorite hunting in Australia is only done by the truest of zealots and the wide-open spaces of the Outback do not attract professional meteorite hunters.

Even though Aussies may not have the most favorable laws for commercial meteorite collection, they, without question, have the best meteorite names in the game. As a rule, official meteorite names are

---

* I know this sounds strange since Australia does not have a king or a queen, so what "crown" are they talking about? I guess British colonialism dies a slower death in some places than in others.

taken from the name of a nearby town or location where the meteorite is found. And if you ever look at the list of meteorites from Australia, you will develop serious envy that folks Down Under get to occasionally give directions to places with names like Billygoat Donga, Cocklebiddy, or Laundry Rockhole. So, next time you are in Australia, I kindly ask you to send me a postcard from any one of these places and I will proudly hang it on my refrigerator.

Best **official**—not a joke—Australian meteorite names
(written in Comic Sans because it just feels right):

Biduna Blowhole	Camel Donga	Dingle Dell	Kybunga	Pigick
Billygoat Donga	Cartoonkana	Gnowangerup	Laundry Rockhole	Snake Bore
Buckleboo	Cocklebiddy	Johnny's Donga	Millbillillie	Starvation Lake
Bunburra Rockhole	Crab Hole	Kittakittaooloo	Mukinbudin	Wonyulgunna

The Atacama

The western coast of South America, particularly northern Chile, hosts the oldest and driest of all the hot deserts on Earth, the Atacama Desert. This desert has persisted in a hyperarid state for at least 3 million years, with some areas mostly arid for the bulk of 200 million years. It is so dry, there are weather stations in the Atacama that have operated for generations that have never reported rain. Whereas this extended period of dry climate does not make the area ideal for growing rice or kumquats, it has preserved a bumper crop of meteorites. The Atacama may not have the advantage of a white limestone bedrock to highlight meteorites like the Nullarbor, but it has had a *really* long time for meteorites to accumulate, and since it so rarely rains, the meteorites that do land are preserved for shockingly long periods. Multiple meteorites found in the Atacama had lain there for more

than 2 million years, an extremely long time for an exposed meteorite to survive on Earth.*

Because of its impressive aridity and long-term geologic stability, the Atacama is a very important meteorite reservoir and holds the title belt for the highest meteorite density of any of the hot deserts. As of early 2021, there were over 1,900 official meteorites from the slender country of Chile, virtually all of them from the Atacama.

Oman

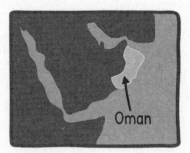

The Sultanate of Oman sits on the southeastern corner of the Arabian Peninsula. Based on this geography and its location on this list, you may have guessed it also boasts an arid climate well suited for meteorite recovery. In addition to having exceptionally flat and dry areas, the Omani desert has areas of intense white terrain. Carbonate beds and salt flats are common in Oman, providing a perfect backdrop to contrast dark meteorites. Private collectors and organized groups took notice of these favorable collection characteristics starting around 1999, and, within ten years, the Omani desert had produced more than 5,000 meteorite fragments. Unlike Australia, meteorite collection is legal in Oman and things can, by the letter of the law, be removed from the country and sold. Many of these fragments have since turned out to be "paired stones"—or pieces of the same parent meteorite found in different locations—which occurs when the meteorite breaks up as it travels

---

* Measurements like these are called "terrestrial residence ages" and are determined by measuring the isotopic compositions of various gases trapped in the samples. These measurements allow scientists to determine how long a sample has been on Earth. Similar methods also make it possible to determine how long a meteorite was flying around space after it was dislodged from its parent body but before it landed on Earth. And yes, you are right, it is awesome that we can figure stuff like that out.

through the atmosphere. Regardless, this is still an incredibly impressive haul. To date, approximately 4,000 distinct *individual* meteorites have been found in Oman. In addition, Oman has been lucky enough to be the landing zones of multiple lunar and Martian meteorites, adding to the scientific and monetary value of the collection zones.

However, before you rush off to find your meteoritic fortunes on the salt planes of the sultanate (or anywhere else), please consider the story of two of the world's best and most experienced meteorite hunters, Michael Farmer and Robert Ward. In 2011, the pair were two weeks into a collecting trip in Oman, Michael's twentieth trip to the sultanate. Having acquired exciting new finds and multiple new fragments of a lunar meteorite that Michael had found on a previous trip in 2005—all in all, it appeared to be quite a successful venture. Successful, that is, until they ended up in the bowels of an Omani prison on charges of illegal mining. Fortunately, the law was on Michael and Robert's side and they were eventually released after appearing before a panel of judges; but that was only after they endured three hellish months, a prison riot, lost forty pounds each, and gained a lifetime of nightmares. Lucky for us, Michael and Robert continue to search for and collect meteorites around the globe.* But no longer in Oman.

The Mighty Sahara

From the above examples, you likely predicted that the world's largest hot desert, the Sahara and its ~3.5 million square miles, would be on a list of productive meteorite collection areas. You would not be wrong. Over the last twenty or so years, the Sahara has become one of the most important areas for meteorite collection and continues to produce

---

* Later that same year, Michael was in Kenya on a collecting trip and, well, I will let him tell it from his 2013 *National Geographic* article about his unique profession . . . "I was down on my knees, with a bag over my head and a machete on my throat and a gun at my head." Luckily, they just robbed him. In another interesting story, Michael smuggled himself into Algeria to look for meteorites and was chased by soldiers for hours through minefields before the Algerian junta chased him back to Morocco.

more samples than can be cataloged and studied year after year, which as a meteorite researcher, is a very good problem to have.

The Sahara encompasses large parts of more than ten gigantic countries, and whereas the political borders do not interrupt the virtually unbroken potential collection zone, the diversity of national laws and sporadic political unrest greatly complicate searches, collection, and export—particularly for nonlocal entities. As such, the vast majority of Saharan meteorites are collected by residents of the desert itself, not by jet-setting professional meteorite hunters or by academic consortiums.

The exact origins of the meteorite boom in the Sahara are difficult to parse, but it seems that a lot of the credit goes to two specific occurrences from 1997. First, a French meteorite collector and dealer named Luc Labenne was tooling around the Sahara in Mauritania with his family when he heard rumors of meteorites being found in neighboring Algeria, so he decided it was worth a look while he was nearby and in a similar region. Within just a few months, Labenne had found more than 200 stones, unambiguously alerting scientists to the potential of the Sahara as a source of meteorites. Later that year, the American meteorite dealer Edwin Thompson established contact with a group of nomads through an English-speaking Moroccan middleman who had tracked and retrieved a sample from a fall in the El Hammami Mountains. Thompson flew to Mauritania in November 1997 and, thanks to help from the nomads and an exceptionally motivating wad of American cash, located over 200 kg of what is now officially known as the El Hammami meteorite. The scientific value of the meteorites that Labenne and Thompson had acquired from Mauritania was minimal, but the potential, and more importantly the market and procedure for meteorite collection, was established, starting a "gold rush" of sorts in the greater Sahara region. Word quickly spread, and ancient nomadic trade routes previously used for things like silk, salt, and gold were now being used to transport likely meteorites via camel caravan by the ton. As the nomads became increasingly adept at discerning meteoritic from terrestrial materials, and their profits grew, they roved deeper and deeper into the desert in search of rocks from space.

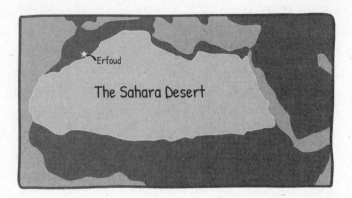

Due to a combination of factors, including the nationality of the original meteoritic middleman, the desert outpost of Erfoud, Morocco,* quickly became the center of the Saharan meteorite trade. However, perhaps most important for this development is the lack of clarity regarding meteorites in Moroccan mining legislation. This legal void has allowed traders from all over North Africa to bring meteorites into Moroccan markets to be sold to international dealers, regardless of their actual country of origin. In just one particular example, many international museums refuse to even consider purchase of meteorites that were found in Algeria due to sticky legal issues of who actually owns the meteorite if it fell on Algerian soil. However, no such issues exist with meteorites found in Morocco. As a consequence, there is a shocking statistical overabundance of meteorites "found" in Morocco by nomads, and a correspondingly and perhaps not-so-shocking lack of meteorites found in countries with more problematic export policies.†

---

* For the billions of Brendan Fraser fans out there, Erfoud and the surrounding desert was where much of the terrible 1999 movie *The Mummy* was filmed.

† There have recently been calls from within Morocco to clarify laws on this subject to ensure that at least some of the material recovered in Morocco stays in the country. This is a delicate issue, because while the idea of establishing museums in Morocco and holding on to meteoritic treasures from the country is a noble pursuit, the livelihoods of tens of thousands of people count on the meteorite trade in the country. Any change in policy could severely impact their lives, not to mention the flow of meteorites to the market and academic institutions for scientific study.

Overall, it is difficult to wrap one's head around the number of meteorites that have come from the Sahara in the last few decades. Around 2,500 samples have known find locations in a variety of countries in the region, and another ~9,000 meteorites have been given the umbrella designation "Northwest Africa," or "NWA" because where the sample was collected is unknown or untold. In 1999, 45 NWA samples were classified. Twenty years later, that number has grown by more than an order of magnitude, with barrels full of samples piling up just waiting to be classified. Ordinary chondrites—the most common type of meteorite finds and falls—are so abundant from desert collection that these types of meteorites are often ignored to save room for more interesting samples that will fetch higher prices. There are even tales of collectors piling up literal tons of ordinary chondrites in the desert to come back to at a later date if it was ever worth the bother.

A Song of Rocks and Ice

The above-mentioned collection areas from hot desert regions have been incredibly important for a lot of reasons and have vastly increased the amount and type of meteoritic material available for study. However, from a sample volume perspective, there is no doubt which area is king: all hail Antarctica.

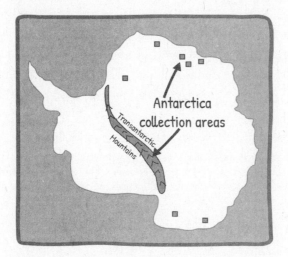

The most basic ingredients required for meteorite collection are (1) a dry area for rocks to land, and (2) people to find said rocks. Antarctica has the first in abundance, but since the continent lacks a permanent population and has less than 5,000 people residing on it at any given time, you would be forgiven if you thought this would be a major problem for Antarctica in the race for meteorite collection supremacy. But Antarctica has a few tricks up its sleeves that hot deserts simply cannot compete with.

The first meteorite discovered in Antarctica was in 1912, and a handful of others were recognized by various expeditions over the next five decades. However, the true potential for meteorite hunting in Antarctica was realized after a Japanese field excursion in December 1969.* The excursion was charged with setting up survey stations in the East Antarctic ice sheet to facilitate the study of glacial movement in the area. In that single campaign—a campaign that was not even designed to be looking for meteorites—the group found nine individual stones of a variety of types. This, importantly, meant that these rocks were not just a single stone that broke up in the atmosphere; they came from multiple meteorite falls from different parent bodies strewn across the glacial ice. When the circumstances of the finds were reported a few years later at the annual meeting of the Meteoritical Society, the late Bill Cassidy, a professor at the University of Pittsburgh at the time, grew very excited about the prospect of meteorites in Antarctica and almost immediately started formulating plans for collection trips to the icy expanse.

Cassidy was not deterred by his perhaps predictable multiple rejections for funding from the National Science Foundation.† Keeping the idea alive through collaboration with colleagues from Japan, it was decided that the same glaciology group that had reported the finds in 1969

---

* It is likely not a coincidence that this was recognized in 1969, the same year of Allende and Murchison meteorite falls and humans landing on the moon. Space was on the brain at the time, and that helped the field of meteoritics an incredible amount.

† Sadly, the reviews were not pretty and are said to have included the term "ludicrous" and had the general tone of wondering why we would even bother collecting meteorites in the first place.

would also search for meteorites during their 1975–76 season. With this additional task of actively looking for samples when they could, that same patch of ice at the base of the Yamoto Mountains yielded 663 more specimens. Not surprisingly, Cassidy was approved shortly thereafter to lead a team of two American scientists out of McMurdo Station to search for meteorites the next season in Antarctica. It is funny what a little international competition can do to loosen up the purse strings.

Since the first 1976–77 dedicated field season, the United States as well as other nations has sent teams every year to collect meteorites from the glaciers of Antarctica. The U.S.-led team, called ANSMET (for the ANtarctic Search for METeorites), generally takes between six to twelve volunteers* to camp and search along the slopes of the Transantarctic Mountains. Each season lasts around six to seven weeks, with the weather largely dictating how much time is spent in a nine-square-foot tent, potentially wishing for a better companion for such tight quarters. But when the weather cooperates, volunteers get to search the blue ice fields or glacial moraines on a snowmobile for primitive pieces of the Solar System for the betterment of science. Even more than the barrels upon barrels of extraterrestrial material recovered from the ice by these intrepid search parties, according to every ANSMET participant I've spoken to, the most memorable thing the group takes home after the field season are the fierce bonds of friendship born from battling and conquering the extreme environment of the Antarctic with their cohorts.

The amount of material that has been collected on trips to Antarctica over the last ~45 years is almost unfathomable. Collectively, teams from across the globe† have found more extraterrestrial material in Antarctica than has been collected over recorded time, everywhere else

---

* This is a highly competitive process and being selected for a trip is considered a huge honor by the meteorite community. Because the team needs experienced people to deal with the dangers and difficulties of living in Antarctica, only a couple "new" people are allowed to go every year. This also means if you have been once, you have a decent chance at being selected again.

† The primary players are the United States and Japan; however, China and various European contingents have also organized collection trips.

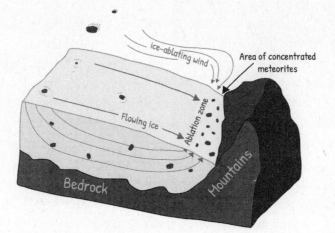

The great meteorite conveyer belt. When meteorites land on top of glaciers, they sink into the ice as they move with the glacier. If the glacier runs into a mountain range, the meteorites pile up with the ice at the base of the mountain range. The ice is slowly ablated away by wind, leaving a high concentration of meteorites.

combined. To date, well over 45,000 officially classified meteorites have been found in Antarctica.

But this insane number of meteorites was not collected by expeditions just looking around the ice randomly. Collection areas are chosen for a variety of reasons, but one of the most important is to look in areas where the downward flow of a glacier meets the base of a mountain. As it turns out, the glaciers of Antarctica are providing an incredible service to the meteorite community, acting as a giant (but slow) conveyer belt of meteorites. Like anywhere on Earth, meteorites can and do land on top of glaciers. Since glaciers are giant rivers of ice slowly creeping downhill, any meteorite that lands on a glacier catches a free ride on the flowing ice. If a glacier runs into a mountain, the ice and all of its passengers pile up at its base. As this happens, dry Antarctic winds ablate the ice as it piles up, exposing the meteorites. When this process is continued over great spans of time, this results in far higher concentrations of meteorites at the base of the mountain than would be on any random stretch of ice, making the bases of the Transantarctic Mountains an absolute wonderland for meteorite collection.

Left: *The team snowmobiling across a frozen expanse looking for pieces of the early Solar System.* Right: ANSMET *member Emilie Dunham with a meteorite found on the ice during the 2019–20 campaign.*

From Meteorite Collection to Sample Distribution

While the meteorites collected during Antarctic expeditions are curated by NASA at the Johnson Space Center and are available for scientific study by request, most meteorites collected outside of Antarctica eventually are dumped into the free market at some point. Some of the less reputable dealers will immediately place stones up for sale as soon as they are found, either in a janky roadside rock and mineral shop or on the Internet. This *can* be a way to buy actual meteorites, but it can also be a way for the uninitiated to greatly overpay for a piece of concrete, iron slag, or petrified dog turd. On the other hand, reputable meteorite hunters and dealers routinely partner with academic institutions to officially classify new meteorites, meaning that once the stone is classified[*]

---

[*] One of the genuinely nice things about meteorites is that they are basically impossible to fake. With a little bit of training, most people can distinguish between meteorites

by an expert and some basic chemical and mineralogical measurements are taken, the stone and its designation are entered into the Meteoritical Bulletin Database maintained by the Meteoritical Society. There are significant benefits to all parties with this arrangement: the dealer gets expert classification information to provide to potential buyers, and the classifying parties keep a portion of the sample in that institution's collection for later study, as well as obtain information from the finder/dealer about where and how it was acquired—information that is generally lost if the meteorite is sold out of a wheelbarrow on the side of the road.

The Tucson Gem and Mineral Show

Regardless of how, where, or if a meteorite was classified, the goal of most dealers and hunters is to sell or trade the sample for something of value to them. You know, capitalism. Of course, these sales and trades happen at a variety of venues, large and small, on the Internet and in person, but by far, the most important place where meteorite deals are made is at the annual Tucson Gem and Mineral Show in Tucson, Arizona. Speaking from experience, the event is absolutely surreal, whether or not you are into rocks. The event is so large, it is not simply held at a convention center, but is spread out across the city of Tucson. Meteorite dealers in particular seem to try to avoid the main convention hall, where enormous mineral displays and spectacular fossils are the prime attractions. Instead, the space rock folks gravitate to hotels like the Days Inn, the Ramada, the Howard Johnson. And I don't mean hotel lobbies or banquet halls, I mean the individual rooms. Many dealers routinely rent a room for the week and spread their merchandise across the floral patterns atop king-size beds. Potential buyers walk from room to room awkwardly navigating narrow passages searching for deals. It sounds shady as hell, but it would

---

and terrestrial rocks in most cases. Counterfeiting rocks of really any type is not a reasonable path to fortune, and the textures and minerals present in many meteorites are essentially impossible to re-create on Earth, even with incredibly sophisticated equipment. It can sometimes be difficult to tell if something *is* definitively a meteorite, but it is usually very easy to tell when something *is not* a meteorite.

not be weird at this show for someone to spend $150,000 in the kitchenette of a Days Inn to legitimately purchase a piece of Mars.

Consequences of the Meteorite Trade

The windfall of meteorites provided over the last few decades from the multitude of collection areas on Earth has been a game changer for meteorite studies. The exponentially higher number of samples has not only furnished a lot more material to study but has also helped researchers recognize new groups of meteorites, better understand the flux of meteorites to Earth, and even investigate them as potential paleoclimate markers tracking desert conditions over geologic time as meteorites degrade. Whereas it would be difficult to argue that the scientific consequences have been anything but positive, it is important to at least consider the potential negatives for society that have developed,

A truly bizarre picture of two boys sitting in the Willamette meteorite, not having any fun at all.

or could develop, through meteorite collection. Such concern, after all, has not always been exercised during the extraction of more traditional geologic resources such as diamonds or oil.

The first, and probably most challenging concern to deal with, is the case of meteorites now part of private or institutional collections that were/are treated as sacred artifacts by Indigenous peoples. There are a few examples of this, perhaps the most famous being the Willamette meteorite, which was found in 1902 in the Willamette River Valley of Oregon, where it has long been sacred to members of the Confederated Tribes of the Grand Ronde Community of Oregon. As you might have guessed, shortly after its existence was discovered by a nonindigenous local resident, the meteorite was transported to the American Museum of Natural History in New York City, where it remains on permanent display.* As these cases are rare, they tend to be, and probably should be, dealt with on an individual basis. However, issues surrounding modern meteorite collection are a different bucket of worms entirely.

Depending on the situation, the effect of modern meteorite collecting can range from being felt by just a handful of people, to altering the economics of entire countries. There are two main ways of finding previously unknown meteorites. The first and most exciting scenario is when a fireball occurs where fresh meteoritic material lands on the ground. It is then important to gather the material as quickly as possible; the longer the sample sits around on Earth, the less pristine it is when studied. From a scientific perspective, gathering material before the sample is rained on is incredibly advantageous. From an economic one, the sample is also worth more on the open market when the reports are fresh in people's minds. This urgency then means when a fall occurs, a handful of outsiders rush to the scene in country X and scour the area for something that they can make a lot of money from. At first glance, this may carry echoes of colonialism. But meteorite hunters are

---

* There is currently an agreement between the Confederated Tribes of the Grand Ronde Community and the American Museum of Natural History, and a small piece of the meteorite has been returned as a symbolic gesture.

certainly not there to steal the rocks and subjugate the landowners; they happily pay the landowners for the stones they find based on negotiation. They will pay less than what they make in the meteorite market, but it is a business, and businesses like this make money by buying something at a lower price than they sell it for. The hunter would make the reasonable argument that if the rock were not located and identified, it would be worth nothing to the landowner, and they are paying them a nice sum of money for something that happened to randomly land on their property just a few days prior. On the other hand, others might argue that the hunter is taking advantage of the less meteorite savvy, making a significant profit from something that should instead result in a hefty payday or family heirloom for the lucky landowner. Both sides have points, but this arrangement generally seems to work out for all parties: the landowner gets paid a sum of money for nothing more than a stroke of cosmic luck, the meteorite hunters obtain new samples to put on the market, and science benefits due to the quick actions and rapid recovery.

The other meteorite collecting scenario is far less urgent and occurs when there was no witnessed fall, but it affects a far greater number of people. In practice, this largely involves meteorite hunters hiking around dry lakebeds, salt flats, or otherwise barren landscapes searching for pieces of the early Solar System in various places around the world. How this is dealt with completely depends on the country, as discussed in previous examples. In the Australian example, thousands of meteorites may end up wasting away in the Outback because there is no monetary incentive to go look for them. As such, nonacademic hunters in general will not spend time searching for things they know will largely be confiscated by the mysterious "Crown" anyway, and science loses out on a lot of potential samples because of it.

A recent example of changing laws takes us to the Campo del Cielo meteorite and strewn field of Argentina. More than 100 tons of Campo del Cielo iron meteorite material has been recovered to date, making it one of the most commonly available iron meteorites in the world. To summarize a complicated story spanning multiple decades, following the finding, unearthing, abandonment, and attempted later removal of a specimen dubbed "El Chaco," a series of provincial Argentinian laws were passed in 1994 to keep pieces of the meteorite in their original locations. It is worth noting, however, that those "original locations" are almost entirely underground, meaning that the likely very significant portion of the remaining material from this meteorite will just rust underground without anyone ever even knowing about it. Authorities have reportedly even sealed into drums and reburied confiscated pieces of the meteorite. While it can easily be argued that legislation written to protect, properly conserve, study, and display natural history specimens such as meteorites benefits society as a whole, nobody benefits from this particular situation: not the locals, not the government, not the environment, not the scientists, and certainly not meteorite hunters or the meteorites themselves.

The opposite extreme is a place like Morocco, where tons upon tons of meteorites legally leave the country every month, and very few, if any, stay in the hands of Moroccan museums or researchers. On its face, that also seems like a bad situation, particularly for Morocco. However, there are a limited number of meteorite researchers in Morocco and few museums equipped to curate any significant number of meteorites. Until the infrastructure and capabilities are better supported in Morocco, keeping significant material in country would not necessarily be good for Morocco, and it would certainly not be scientifically prudent, as potentially exciting samples would languish unstudied.

Given humanity's history of natural resource exploitation in developing nations, it is understandable to think of the exploding business of meteorite collection in this context. And it is easy to present a few individual examples as imperfect situations in the worldwide meteorite market. However, it is also reasonably easy to argue that the status quo for meteorite recovery is overall pretty fair and robust—the system in place works surprisingly well for most people involved, from landowner to hunter to scientist. Additionally, there is very little environmental impact associated with meteorite removal. To say there are no issues is folly, but the general system has developed in a way that seems to maximize scientific output and minimize the negative impacts, which is a good balance to shoot for.

Regardless of the discussion about the morality and business of meteorites, meteorite monetization has been both a blessing and a curse for meteorite researchers. On the good side of things, it has produced a market in which people go searching for meteorites thinking a large payday might await them if they find the right kind of stone. On the bad side of things, the open market can be a difficult place to compete for samples if your research group is working on a tight budget. Scholarly-leaning hunters and collectors, museums, and the Antarctic meteorite collections are what keep research affordable and possible.

Overall, the greatly increased availability of samples in the recent decades has been a huge boon for understanding our origins and the early Solar System. However, questions from the public regarding the financial worth of meteoritic samples are constant reminders that meteoriticists must continue to discuss the importance of what we learn from the samples and stress that these rocks are not just another fungible commodity; they are unique time capsules of information from the deep past that contain information we have no other way of obtaining.

Eight

Meteorite Mischief and Mitigation

Do you have an impending sense of doom that you will be unlucky enough to be struck by a meteorite? Or, even more consequentially, do you think Earth as we know it will end with the supersonic arrival of a giant flying rock from space? The effect of Hollywood disaster movies on the human psyche is hard to overstate here, but as you may have recently read, like, perhaps in all previous chapters of this book, space-based material landing on Earth is a thing that happens. And if it hits you or it is big, it will quite likely be problematic. As such, addressing the dangers and annoyances associated with flying bits of space rock is probably worth doing, because—and I cannot stress this enough here—♫ I dooonnn't wanna miss a thiiiiing. ♫

Historically, people have not feared death by meteorite. The main reasons for this are (1) only in the last few hundred years did most people become aware that rocks could fall from space, and (2) past generations had other, more pressing things to worry about (for example, avoiding cholera, famine, and/or oxen stampedes). However, modern humans widely understand that they *could* die from a flying rock and many of us have loads of extra time to worry about such things. This combination, and the fact that stray meteorites are far more exciting to worry about than realistic issues like heart disease, car accidents, pandemics, or gun violence, means there has been a significant uptick in the death-by-meteorite fear department, which is why it feels prudent to address it here.

Numerous stories *suggest* that meteorites have hit people,[*] and many of these may be true; but, unless the fall is witnessed by other people, recovered, and analyzed, the authenticity of such events can be difficult to parse, particularly given the financial incentives now realized for possessing a physical piece of such a bizarre event or the short-lived notoriety associated with human/meteorite interaction. This is certainly not meant to disregard other stories of meteorite strikes on people, but confirming such events is not straightforward.

Certainly, an interesting addition to this book might be a list of the hundreds of deadly meteorite strikes, narrow escapes, and associated harrowing stories, but thankfully, this is not possible. Even generously, historical evidence of death by meteorites is thin. However, there is solid evidence of an unfortunately positioned cow in northwestern Venezuela that met an untimely end thanks to a direct hit from the Valero meteorite[†] on October 15, 1972. After seeing an intense flash of light and hearing a loud bang in the evening, three farm hands went out the next morning to investigate and found that one of their bovine buddies had been in the direct path of a ~110 lb. (50 kg) falling rock. It did not end well for the cow, but the farmer did report that what was left of the cow was butchered and eaten. Moooo-ving on.

The best-documented case of a meteorite impacting a human comes from near the small town of Sylacauga, Alabama, in 1954. On November 30, Ann Hodges was enjoying an afternoon nap on the couch when a grapefruit-size meteorite crashed through the roof of her home,

---

[*] A list of many of the confirmed and likely events can be found in the *International Comet Quarterly,* maintained by the Earth and Planetary Sciences Department at Harvard University.

[†] This is not to be confused with the massive ~3.8-ton Vaca Muerta meteorite, a meteorite found in the "Vaca Muerta" region of Chile, which translates to "dead cow." This is incredibly confusing, but the Vaca Muerta meteorite almost certainly did not kill a cow, and the Valero meteorite almost certainly did. Clear?

ricocheted off a large wooden radio, and struck her on the hip. Even though it caused some serious bruising, things could have been worse for Hodges—she managed to survive a meteorite hitting her, and she became a minor celebrity in the process,* which is far more productive than most of my afternoon naps.

A more recent case of a person being struck by a meteorite comes from Mbale, Uganda, in 1992. The rather large Mbale meteorite exploded in the sky during its journey through Earth's atmosphere, causing a loud noise that was heard widely throughout the heavily populated area. The resulting meteorite shower covered over eight square miles and resulted in over 330 pounds (150 kg) of material being recovered, mostly by local military personnel.† While many buildings were pep-

Meteo-rite place, meteo-wrong time. The hole in the Hodges house caused by the Sylacauga meteorite in 1954.

---

* The celebrity was short-lived, and the events of Hodges's life after being struck were not necessarily all positive, causing some people to think the meteorite was a bit of a curse for her and her family.

† As discussed in Chapter 3, a sizable portion of the Mbale meteorite was ground up and used as a hopeful cure for AIDS, which was rampant in the region at the time.

pered and many falling stones were observed, the only strike victim was a young boy who was hit, luckily, by only a very small piece of the meteorite after it had bounced off a tree.

Digging through historical archives, it is at least possible to get a sense of meteorite strikes and their direct effect on humans through the generations, even though it is almost impossible to confirm these cases so long after the fact. One very impressive list was compiled by researcher Kevin Yau and colleagues by looking through ~2,600 years of reported Chinese meteorite falls that have impacted humans or human structures. Yau came up with seven events that resulted in human casualties, some of which were reported to have killed tens of thousands of people, although there are good reasons to believe the absolute numbers from the larger events are unreliable. Just uncovered in 2020 were three corroborating government documents written in Ottoman Turkish that discuss the death of one man and the serious injury of another from a meteoritic airburst that occurred in what is now modern-day Iraq on August 22, 1888. It is hard to argue that this is not a legitimate record of a human death caused by a meteorite, but the offending stone has yet to be located and verified. There are other reported, and undoubtedly many more unreported, fascinating events such as these reports scattered throughout history, but most historians who keep track of such things likely would claim they have better things to do than comb through generation after generation of governmental records in various languages to find evidence of a truly obscure way of dying. Understandable. But even if we are extremely charitable to the records, known and unknown, and say that 50,000 people in all of history have died a meteorite-related death, then this would still be an incredibly rare way to die when statistics are considered. Fifty thousand people killed by meteorites would still be less than the number of people in the last fifty years that have died by falling off ladders.

The Threat Is Still Real

Even though deaths caused by meteorite strikes are few, there have still been consequential events over the years that have raised some serious

eyebrows when it comes to the potential dangers of rocks from space.*
The first serious wake-up call to people and governments around the
world happened in 1908, in what is almost the geographic center of Russia. The Tunguska Event, as it is known from the name of a nearby river,
is the largest "impact" event in recorded human history, although the
event did not result in the formation of a traditional crater. The exact
nature of the event is still unknown, but scientists largely agree that the
massive explosion was caused by an airburst from a meteorite or comet.
Seismic readings from the event were noted across Europe and Asia,
and airwaves from the blast were detected on the East Coast of North
America. The destruction caused by the explosion is almost unimaginable: it leveled almost a hundred million trees over ~830 square miles,
an area larger than twenty-five of the countries on Earth. It would have
destroyed any major metropolitan area, and deaths could have been in

*Downed trees remained as scars when scientific expeditions were studying the
Tunguska Event about twenty years later.*

---

* Just ask the dinosaurs.

the millions. Fortunately, the event happened in an incredibly sparsely populated area of Siberia where only three people were reportedly killed, along with a herd of reindeer. Just like for Ann Hodges, things could have been much worse for humanity.

Due to the remoteness of the location, proper study of the Tunguska Event was limited, and scientific expeditions did not reach the site until 1921, thirteen years after the event occurred. Damage from meteoritic material, particularly significant damage, is rare, so the fact that the Tunguska Event did not spur much action may not be surprising given its limited financial and human toll—and it was 1908, so technology was a bit more limited by today's standards. But Tunguska at least put the planet on notice of the dangers that lurk in deep space.

Small Potatoes (and Similarly Sized Rocks)

There is nothing to compare with Tunguska in terms of damage, but there are numerous amusing stories from around the globe, like the poor mailbox in Claxton, Georgia, that got the meteorite treatment in 1984, and the Madrid meteorite that fell just outside the steps to the Natural History Museum and was collected by the director of the museum's Meteorological Institute. There was even a golf-ball-size meteorite that lodged itself into the home of a French family with the last name of Comette in 2011. And, perhaps most famously, Peekskill, New York, was the site in 1992 where a meteorite violently found its way through the trunk of a cherry red 1980 Chevy Malibu. This may have seemed like incredibly bad luck to eighteen-year-old Michelle Knapp, who had just recently purchased the car for four hundred dollars, but time may have convinced her this was a blessing in disguise. In 2012, the broken rear taillight bulb and original car title sold for over $5,000, and the 26 lb. (12 kg) meteorite currently sells for >$150 per gram on the open market. People love a rock with a story, apparently.

One could go on for pages and pages about stories from meteorite falls throughout the years—and they are all interesting in their own right and chronicled in a number of places—but these were all inconsequential falls on a global scale. None of these meteorite impacts caused

Me posing in front of the famous Malibu while it was on display at a meteorite conference in Paris in 2018. In taking the picture, my lovely wife somehow managed to position herself and the corner support to block almost the entire damaged area of the car. I may never forgive her for it.

much global consternation or political dialogue about protecting our planet from flying rocks, but should we be more concerned about the dangers from above?

The First Step Is Realizing You Have a Problem

When scientists showed in the 1980s that the dinosaurs were eradicated by a massive rock from space, people started to get a bit concerned that our species could also go extinct in the blink of an eye. The U.S. Congress mandated to NASA in 1992 that they should locate 90 percent of the near-Earth asteroids larger than one kilometer within ten years. The perils of planetary pinball got a bit more attention still when the comet Shoemaker-Levy 9 slammed into Jupiter in 1994, releasing an amount of energy estimated at ~600 times the world's entire nuclear arsenal,* leaving a truly massive scar on a truly massive planet. But it was not like we were actively developing a plan in case something similar was detected headed our way. The pair of 1998 movies *Deep Impact* and *Armageddon*

---

* Just for some perspective, the world's nuclear arsenal at the time contained over 38,000 weapons, most of them many times larger than those used at the end of World War II.

were potentially helpful to raise the issue that we might want to think about actively forming a mitigation plan in case we found ourselves in the crosshairs, but these were just movies from Hollywood, so how serious should politicians take this idea? Luckily, scientists take it quite seriously and have been combing the skies for a while now, searching for what are known as near-Earth objects (NEOs) in an attempt to identify objects that might be on a collision course with our home planet. Many groups at universities and governmental agencies are dedicated to looking for particular sizes and types of NEOs that might become problematic, but the collective monitoring system is termed Space-guard* (awesome name) and has been scanning the skies as an affiliated group since the mid-1990s.

The issue got another major shot in the arm on February 15, 2013, when the Siberian sky opened (again) for the arrival of the Chelyabinsk meteorite. At 9:20 in the morning, the meteor lit up a large area of the sky brighter than the Sun as the object careened toward Earth. The meteor exploded before it reached the surface, causing a massive airburst

A photo of the Chelyabinsk meteor.

---

* This is named after a fictional space-monitoring system in the Arthur C. Clarke science fiction book *Rendezvous with Rama*.

that shattered windows and damaged thousands of buildings in the area. This airburst caused far more damage than the surviving flying chunks of rock that then became the Chelyabinsk meteorite. Even though the area is a relatively sparsely populated part of Russia, the event still injured more than a thousand people and caused well over a billion rubles (>$33 million) in structural damage. The asteroid that caused the event, which had an estimated diameter of ~66 ft (20 m), was not detected until it entered the atmosphere.

Just some of the damage the airburst caused to one local building.

The day of the incident, the rarely-heard-from United Nations Office for Outer Space Affairs suggested creating an "Action Team on Near-Earth Objects," and the U.S. House Science Committee met shortly after to discuss how to deal with threats from space, although it is possible the U.S. response was also partially because Congress gets extra nervous when Russia blows anything up, deliberately or not. Either way, following Chelyabinsk, NASA announced the Planetary Defense Coordination Office, whose mission it is to use "applied planetary science" to address the NEO impact hazard. The office is tasked with finding and tracking all potentially hazardous objects larger than 30–50 meters (about twice the size of the Chelyabinsk offender) and to develop and coordinate strategies and technologies for mitigating any potential impactors to Earth. So, yes, it is basically a scientific version of the movie *Armageddon* with all options on the table.

Possible Mitigation Strategies*

So, what would we do if scientists located a large asteroid or comet headed our way? After instantaneously pooping our collective pants, there are two types of realistic strategies that can basically be classified into (1) gradual methods and (2) sudden methods, where both methods are just as they sound. The gradual approach slowly changes the path of the object over time, diverting it from an Earth-crossing orbit. The sudden approach, well, blows the bastard up. Intelligently choosing between these options requires that we know a decent amount about the impending collider, such as size and shape, speed, and basic composition. Also important is the predicted time before it hits Earth; if we have multiple years to decades, we have a lot more options. If we have months to a year, well, sudden methods will be the only game in town. Either way, we also *really* do not want to make things worse—having accurate measurements and good data are key to preventing us from pulling a rock into our orbital path that was not going to hit us in the first place or blowing up a rock in which twenty medium-size pieces hit us instead of one large rock missing us.

Let us assume we have detected a species-ending rock that we are sure is going to hit us in ten years. How do you slowly change the path

---

* Natalie Starkey does a very nice job of describing a variety of mitigation strategies in Chapter 10 of her book *Catching Stardust* if you would like additional ideas or more detail about some of the ideas discussed here.

of a giant flying rock in space? The first step is to get a spaceship up near the object. After that, the most passive approach is to let gravity do most of the work for you. Since gravity acts like an invisible tether, the object is gravitationally attracted to the spacecraft once they are near one another. As such, thanks to gravity, you can slowly change the course of the spacecraft with gentle thruster bumps every now and again, ever so slightly dragging the rock along with you in space—this method is given the amazing name of the "gravity tractor." Other, more active options involve the spacecraft firing a constant stream of ions at the object to slowly divert it or actively covering the NEO in some sort of white paint to change how it responds to solar radiation, ultimately resulting in a path adjustment. But these options all require years of preparation and years of action in space. What if we only have one year or six months before impact? Well, they most definitely did not make a movie about painting the side of an asteroid white.

If you are as big of a fan of Atari video games from the 1980s as I am, you may imagine the active engagement strategies similar to either the game Missile Command, where the dangers come from rocks instead of missiles and they are targeted from ground-based defenses, or, of course, the classic Asteroids, where you fly around in a spaceship and try to destroy or alter the path of the troublesome rock. Well, congratulate yourself (and Atari), because that is about right. If Earth has a short time to act before we are obliterated, we fire something from the ground to hit the object and break it up, we fly up there to shoot at it, or alternatively, we modify the gravity tractor into a real tractor and drag the asteroid/comet out of harm's way.

This, of course, sounds like engineering fantasy, but lucky for us, planetary scientists are already hard at work testing some of these diversion techniques. In 2005, NASA successfully launched a small projectile and hit the comet Tempel 1 as it whizzed around the Solar System in the Deep Impact mission. The mission helped scientists learn a ton about the composition of the comet as well as acting as a proof-of-concept in case we must do something similar in the future to save our planet from destruction. The joint mission between NASA and the European Space Agency (ESA) called DART (Double Asteroid Redirection Test)

If we don't have a lot of time

If we detect a large rock coming our way, the type of defense we mount will depend on how much time we have to act.

acts as follow-up on the Deep Impact mission, only on a much larger scale, with the goal to measurably alter the path of the binary asteroid Didymos. Other technologies that actively drag asteroids out of their current path are in the planning stages and will likely be tested in the coming decade(s).*

It is unclear which, if any, of these planetary defense strategies will be needed while humans inhabit Earth. But it is an almost statistical certainty that our planet will be struck by a large asteroid or comet at some point in the future; it is just a matter of when, and if we will be around to care. If it is during our time on Earth, it would be nice to have the opportunity to fight back. And unless we invest in the detection capabilities, experiments, and engineering expertise that currently sound like science fiction, then we will be helpless to protect ourselves from the statistics of the Universe. And if the possibility of saving all inhabitants on Earth by actively avoiding collision with a large asteroid does not qualify as a good reason to study space science, then there is little that will change your mind.

---

* This technology is being heavily supported by many newly formed space mining groups in hopes they someday will be able to drag an asteroid full of scarce metals into Earth's orbit to make it easier to exploit.

Nine

Modern Meteorite Research

The previous two-plus centuries of scientific research on meteorites opened our eyes to numerous incredible discoveries, from finding star fossils older than our Solar System to helping us understand how planets form and evolve; but do we have it all figured out? Not even close. Through the study of meteorites, we may have already learned an immense amount about our stellar environment, our natural world, and the world constructed by humanity, but numerous avenues of meteoritic research are still tackling some very big and very interesting questions. How did the Solar System start? We don't really know. Has life ever existed on Mars? Not sure. What types of environments are required to produce complex organic molecules? Good question. The exciting parts of life occur when you don't know what is going to happen, so I want to close the book on the fundamental thing I find most thrilling about science in general: there are a lot of fascinating questions that we don't yet have answers to—so we search for them. Below is by no means an exhaustive collection of the exciting things happening in meteoritics, but a list of the ongoing research questions in the field of space science that I personally find particularly compelling, and briefly how they are being answered.

Tracking the Fall of Meteorites–
Technology to the Rescue

People often ask where meteorites come from. Every so often we get a chunk of the Moon or Mars, but the primary answer is the asteroid belt. But the asteroid belt is ~93 million miles (150 million km) wide, contains ~1 million asteroids that are greater than one kilometer in diameter, and perhaps billions of smaller objects. So, the answer of "the asteroid belt," while correct, is incredibly vague. We can do better.

On April 7, 1959, the fall of the Příbram meteorite in what was then Czechoslovakia was captured on film by multiple cameras at the Ondřejov Observatory—the first time this had happened. Capturing a fireball on film may seem like a cool but strange thing to want to do, but the benefit of photographing a fall is huge, at least in theory. First, if you are using a camera with a known shutter speed and you get multiple shots of the fireball, you can determine the meteor's velocity. Second, if you capture the fall with multiple cameras, you can triangulate both the meteor's starting and ending points. This means that if you could find the terminus of where the fireball was pointing, you could find a fresh meteorite, and a fresh meteorite is way more scientifically useful than one that has been lying around for a few thousand years. Additionally, triangulation permits back-calculation of where the sample's orbit originated, meaning a far more precise answer to the question "where do meteorites come from" all the while giving us a much better handle on current Solar System activity.

After the film from the observatory was developed, researchers were able to put all those ideas into practice and mathematically re-create the Příbram fireball's terminus, and within a week the team located almost 13 pounds (5.8 kg) of the meteorite in the predicted area. They also learned its speed and calculated the general orbit of the sample when it was still cruising around space. Those are some seriously cool photography and math skills.

A handful of subsequent meteorite falls have been captured on camera with their trajectories triangulated, but advancements in digital photography and pattern recognition have made cool studies like this infinitely more feasible on the large scale. In 2005, using three remotely operated cameras spaced around the desert, Curtain University in Perth, Australia, set out to observe rocks falling from space in the wide-open landscape of Western Australia.

The first major success using the automated camera network came on July 20, 2007, when a fireball was observed over southwestern Australia. After the highest-probability fall location was calculated, the team recovered—within 100 meters of the predicted location—multiple frag-

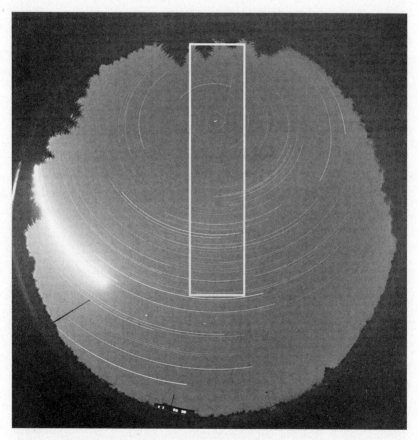

A picture of a fireball crossing star trails over Europe taken with an all-sky camera (highlighted with the box). The segmented image of the fireball not only makes it stand out, but because the camera shutter operates at a known rate, the velocity of the object can be determined.

ments of the meteorite known as Bunburra Rockhole.* This not only proved the remote tracking system worked, but the team recovered a relatively rare type of meteorite shortly after it fell, learned its original location in the asteroid belt, and determined that the rock came very close to hitting Venus in 2001 on its tour of the Solar System prior to landing Down Under.

Since the fall of Bunburra Rockhole, the remote camera network has expanded from three to more than fifty fully autonomous cameras just in Australia, and multiple countries around the globe now watch an impressive amount of sky every night for new additions to our collection and a better understanding of where these rocks are coming from in the first place. But perhaps even more exciting, the project has also started tapping into, by far, the largest camera network on Earth—smartphones. The app "Fireballs in the Sky" allows anyone to get involved in meteorite tracking by using app-reported sightings and locations to assist in tracking meteorite trajectories anywhere in the world. Tracking meteorites with cameras may not be the most groundbreaking science in the world—this was done first in 1959—but it is an absolutely fantastic way to get more people involved and excited about space science, at the same time leading to more and fresher meteorites being collected.

Meteorites and Anthropology

The Bronze Age, the bustling period following the Stone Age, saw humanity grow in remarkable ways. In fact, this age boasts several key human advancements, such as the invention of the wheel, development of writing, the start of widespread farming practices, creation of urban areas, the first system of laws, and the development of extensive trading networks across the globe; all occurred during the Bronze Age. But germane is the reason it was called the Bronze Age and actually relevant for meteoritics; humanity had figured out how to make and work bronze, but it had not yet figured out how to smelt iron—something

---

* I cannot stress enough how much I love Australian meteorite names.

found in abundance in meteorites. As we discussed in Chapter 3, this metallurgical hurdle meant that the only source of iron metal for most cultures was (1) from meteorites, (2) obtained by trading with more technologically advanced societies, or (3) making a local technological breakthrough ushering in a new era for that culture. These options are vastly different anthropologically and knowing the difference is vital to understanding the culture at hand. The first step in untangling this web is figuring out whether the artifact contained metal from a meteorite or not.

Numerous artifacts in museums spanning the globe fit into this category of what is known as "Bronze Age Iron," and anthropologists would greatly benefit from knowing the source of these silvery sojourners. And while determining the difference between meteoritic metal and terrestrial metal is straightforward using modern mass spectrometers, museum curators are understandably reluctant to allow any destructive technique to be used on an irreplaceable artifact, regardless of how little material is needed.

Luckily, this issue has begun to be addressed by a recently developed, nondestructive technique called portable X-ray fluorescence. This technique allows researchers to analyze the elements present in the metal artifact essentially by pointing a fancy X-ray gun at the object and measuring the secondary X-rays that are created. Critically, the metal ratios in the object—for instance, the ratios of iron to cobalt to nickel—can reliably differentiate terrestrial from extraterrestrial sources of iron. Since such nondestructive techniques have a far higher likelihood of being used on precious museum artifacts, we will likely soon be learning a good deal more about the trade habits and technological advancements of ancient cultures.

Life on Mars?

The great David Bowie probably expressed it best through song, but it is undeniable that *lots* of people are interested in the possibility of life on Mars. It is impossible to predict how finding life on another planet would change society, but it is sound to speculate that it would be a pretty big deal. Whether or not we ever find such evidence, meteorites certainly have played, and likely will continue to play, a major role in the search for life on Mars.

As we covered in Chapter 6, the current climate of Mars is a bit on the harsh side. Surface temperatures can be equivalent to a pleasant fall day in New York City, but they can also drop down to –240°F (–153°C). The average surface temperature of Mars is a less-than-balmy –81°F (–63°C), compared with 59°F (15°C) here on Earth. And perhaps even more important than the chilly temps, Mars's thin atmosphere does not block life-devastating ultraviolet (UV) radiation, so extant life on the surface of the planet is highly unlikely. However, while it is unlikely that life *currently* exists on the Red Planet, recent research has uncovered the planet used to be far more hospitable. Mars had a substantially thicker atmosphere, as well as moving water, lakes, and maybe even a shave-ice stand at some point in its past, making it reasonable to speculate Mars could have hosted life. But what would indicate evidence of past life on Mars, keeping in mind the missteps of previous claims of what *looked* to be fossilized microbes?

One of the most important discoveries from the study of Mars in the last few decades is the overwhelming evidence that liquid water once existed on its surface. For example, researchers have found on Mars clay minerals that form in wet environments, like the bottom of a shallow sea or a lake. Importantly, many of these clay-rich areas have been left undisturbed for millions upon millions of years, meaning if they did contain life in whatever form, traces of it should still be there to find (here's looking at you, *Perseverance*). These clay-rich areas are prime targets for upcoming Mars sample return missions, and samples from these ancient water-laden deposits of Mars may even make it to Earth's laboratories for detailed study before I retire. What does this have to do with meteorites? If you are looking for evidence of past life on Mars, it is critical to know as much as possible about the environments on Mars over time, and a lot of this information comes from meteorites. For example, many Martian meteorites contain clay and other potentially biologically important phases that can form without interaction with any organisms, and once newly collected targeted samples that *might* have hosted life in the past arrive on Earth, work done on Martian meteorite clays will be critical for comparing geologic signatures and potential signatures of life. Additionally, the very same people who develop specialized techniques for microscopic imaging and invent instruments to measure chemical abundances and isotopic ratios that help detect life's signatures will be the researchers who are best prepared to study returned samples of Mars.* The scientific techniques used to study potential signatures of life in meteorites from Mars and samples from the ancient Earth are continuously being honed, and experts are gaining valuable experience to study difficult-to-interpret samples and improving equipment to do the job. This type of work has already taught us an amazing amount about life on Earth and processes on other planets, but there is little point in bringing samples back from Mars if we are not prepared to properly measure and interpret them. Did life ever exist on Mars? It is possible we may never know the answer, but we will certainly never know the answer if we do not search for it.

Organic Molecules—What Came from Space and How Do They Form?

For me, the realization that organic molecules exist at all in meteorites is mind-bending enough, but the fact that such a complex and highly diverse suite of molecules—including life-essential things like sugars, alcohols, and amino acids—exist in abundance in many kinds of meteorites is almost incomprehensibly thought provoking. Researchers continue to identify and isolate organic molecules in meteorites at a dizzying pace. This itself is a fascinating area of ongoing study, as many of these organic molecules are unstudied since they do not participate in the life cycle here on Earth. Of the various types of molecules, amino acids, the backbones of all proteins and enzymes for life on Earth, re-

---

* Nobody can just step up and crank a home run off an MLB pitcher unless they have a lot of practice. You start against Little League pitchers, work your way up through college or the minor leagues until, at some point, you are ready for Mars sample return.

ceive extra attention because they could also be potential precursors to life here and on other planets. Can amino acids form in isolation in deep space conditions, or do these more complicated molecules only form on the assembled asteroids, or the "parent body"? What conditions are required for their preservation, and how do certain conditions alter the molecules once they form?

Modern research shows that amino acid formation and preservation strongly depends on several factors. By systematically studying organic molecules in many types of meteorites that have experienced different temperatures and water activities, researchers are starting to understand potential preservation requirements, but they are also finding that the mineralogy of the parent body is also important.* How these factors—which are slightly different on every meteorite parent body—impact the structure, formation, and preservation of amino acids is just starting to be understood. Based on the diversity of meteoritic amino acids that have been identified, there were likely multiple formation mechanisms at play in the early Solar System and on meteorite parent bodies. But what were these mechanisms? We are still working on that.

Another branch of organic research involves studying why meteorites predominantly contain the "left-handed" varieties of amino acids (the variety life on Earth uses) and whether this helped usher in the origin of life. Were these left-handed excesses created when the amino acids were originally formed, or did they develop during the extended time on the parent body by long-term exposure to certain mineralogical and environmental conditions? We don't know yet, but laboratory experiments that mimic conditions found in space and on meteorite parent bodies could help shed light on this. Additionally, computational models of molecular structures and behavior will likely play a significant role in increasing our understanding of these exciting topics. We

---

* The Antarctica meteorite hunting trips are particularly important for learning about extraterrestrial organic molecules. Antarctic meteorites are generally very well preserved since the samples have not experienced high temperatures or organic contamination here on Earth.

don't yet have a firm grip on exactly how organic molecules are created in space or how they are preserved/altered in meteorites. But we know organic molecules are there in abundance, and we continue to search for the hows and whys while preparing for the implications that come along with those answers. And if you are interested in the origin of organic molecules, there are two space missions that are going to address some important questions in this fascinating field.

Space Missions–Going to the Meteorites

It may seem unfair to count some of the ongoing and upcoming space missions as meteoritic research, but what is building a ship to bring back samples of an asteroid if it is not traveling to the source of the meteorites before they come to us? Going to pick it up ourselves circumvents the "luck" aspect of getting samples we might want from a particular asteroid and it eliminates any worries researchers might have about a meteorite being contaminated or altered during its time on Earth, no matter how short that time might be. And, just like with Mars sample return missions, the people experienced at looking at meteorites will be the same people looking at samples returned from an asteroid, so missions like these are incredibly relevant to the meteorite community.

Sample return missions from places other than the Moon are rare, but they have happened in the past. In 2004, NASA semi-successfully returned a collection of solar wind samples when the Genesis mission crash landed in the Utah desert. Two years later, the Stardust mission retrieved particles from the comet Wild 2, upending our understanding about the contents and formation of comets. In 2010, the Japanese mission Hayabusa returned a few grains of material from the asteroid designated 25143 Itokawa; for a sequel, JAXA launched Hayabusa2 in December 2014. The goal of Hayabusa2 was more significant sample return from a more scientifically interesting asteroid, one designated 162173 Ryugu. The return of material from Ryugu was accomplished on December 9, 2020, with the exciting successful capsule recovery. At the time of writing, scientists are currently carefully unpacking and sorting through the returned treasures in order to study them in detail, free from potential contamination from earthly influences.

Akin to Hayabusa2, and with an agreement to share samples between the agencies, NASA's mission OSIRIS-REx launched in September 2016 on its way to the asteroid 101955 Bennu. This particular asteroid—scientifically similar to Ryugu of Hayabusa2—was selected for OSIRIS-REx because remote observations showed that it is highly enriched in carbonaceous material, very similar to the carbonaceous meteorites that contain abundant organic molecules. Following more than one and a half years of orbiting, studying, and choosing the best site, OSIRIS-REx will sample and return ~60 grams or more of Bennu's pristine carbonaceous material to Earth for detailed scientific scrutiny, with sample return expected September 24, 2023. If the plans for both missions go smoothly, the extraterrestrial organic-rich payloads of Hayabusa2 and OSIRIS-REx will ultimately provide vital information about the source of organic compounds that led to the formation of life on Earth. Not a shabby scientific contribution.

Planetary Arrangement–Are We Special?

One compelling area of current research that combines astronomical and meteoritic study is understanding our Solar System's planetary arrangement. We have known since the invention of the telescope that our Solar System has the small, rocky planets near the Sun, and the "gas giants" such as Jupiter and Saturn farther away. But the recent discovery surge of extra-solar planets from our astronomy friends shows us that this organization is essentially unseen in other stellar systems. Most other stellar systems have gas giants very close to their central star,* essentially swapping Jupiter, Saturn, and Neptune for Mercury, Venus, and Earth. Why is our Solar System in a totally different arrange-

---

* It is also worth noting that many stellar systems are binary stars. The famous "double sunset" of Tatooine depicted in *Star Wars* might not be that uncommon for other worlds.

ment to most others, and was that arrangement consequential for the development of life in the Solar System? If we see a similar arrangement to ours in a different stellar system, is there a higher chance that life may have developed there?

This is of course a multifaceted issue that requires a range of experts in various fields, but meteorites can help contribute to this in an important way. We can use meteorites to reconstruct the primordial architecture of the Solar System, basically to "re-map" it to what it used to look like. As we talked about in Chapter 1, the structure of the Solar System has changed a lot, particularly in its early days. The reason is that a young disk of gas and dust only has so much stuff to go around to build planets: Jupiter and Saturn formed early and sucked up a lot of the mass that was in the disk, essentially leaving scraps for the inner Solar System. While there was enough leftover material for some smaller planets to form, the gas giants continued to grow, and gravity moved these behemoths around to find the most stable orbits with the Sun and one another, causing havoc for the runts trying to eke out a planet near the Sun.

This is where meteorites come in. The smaller planetary bodies that were trying to form bigger planets were scattered throughout the Solar System, from very near the Sun to the outer reaches beyond the gas giants—but now they all live in the same neighborhood, the asteroid belt. Because the meteorites we get come from the asteroid belt when these planetary fails bump into one another and break off small bits, our meteorite collection represents material that originated throughout the Solar System, not just where it currently hangs out. And it would be helpful to understand the formation and evolution of the Solar System (and really cool) to know where all of that material was when it first formed, not just where it ended up after the big reshuffling.

It turns out that meteorites have specific characteristics that can tell us where they originally formed. One very intuitive characteristic, but one of the least helpful for locating their source, is the water content of a meteorite. Water is incredibly common in space but mostly in its solid form—ice. And regardless whether water is liquid or solid, it doesn't

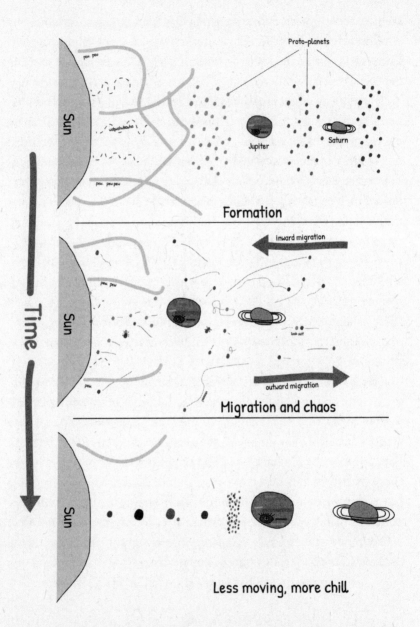

A look back at the figure shown in Chapter 1 about the early disruption in the Solar System caused by the migration of the gas giants. It is possible using the isotopic signatures of meteorites to reconstruct the architecture prior to these migrations, and therefore what our Solar System looked like at its inception.

take too much heat to drive it out and off of a rock. As such, if a body is forming close to a super-hot object like the Sun, it would not have a lot of water because the Sun is hot, and water doesn't stick around when it is hot.* Conversely, if a body is forming far away from its star, it will have a lot of water, since none of it would be driven off. This principle holds in general for the Solar System; objects like comets, which we are fairly sure formed really far out in the Solar System, are approaching 100 percent water-ice, and places like Mercury, which we think formed very close to the Sun, have very little water. The problem with using water content to "cosmolocate" where objects originally formed in the Solar System is that it is easy to change the content of water, even on a large body. If a comet runs into a planet, it can add oceans' worth of water. If a dry asteroid runs into a planet, the heat from impact can boil away any water that was there. So, we need tools other than water content to figure out where planetary bodies formed originally.

One of these such tools that is just starting to come online is something called "paleomagnetism," which is essentially looking at a magnetic field recorded long ago. When the Sun was just getting started, it had an intense magnetic field. Much like magnetizing a nail with a bar magnet, the Sun imparted that magnetic field on objects around it detectable by how different magnetic minerals are aligned. The closer the object was to the young Sun, the more intense the magnetic field that was recorded in the rocks. Unfortunately, this technique does not work on all types of samples and can be compromised if the rocks are not handled appropriately on Earth, but it does provide one way of quantifying how far an object was away from the Sun at the start of the Solar System.

Another such method of determining the distance from the Sun that a sample formed comes from our old friends, isotopes.† With the pre-

---

* If a body has a planetary atmosphere, this can trap water vapor from leaving the body. For a variety of reasons, some bodies have atmospheres, some do not, so this complicates things further.

† More on isotopes and their many uses in the appendix.

cision achievable by modern mass spectrometry, several recent studies have concluded that meteorites record a gradient in their isotopic compositions related to how far out they formed. This may be caused by heat in the inner Solar System able to evaporate certain phases, where the lower temperatures in the outer Solar System means that these phases were not affected in the outer portions of the disk (similar to what happens with ice, only with higher temperatures), or it may have something to do with how material was mixed into our cosmic blender, the spinning disk that became the Sun and planets.

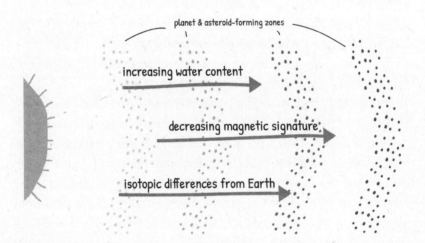

The research of cosmolocation: figuring out where something formed by looking at different properties in a meteorite.

Whether scientists look at water content, isotopic signatures, leftover magnetics, or ideally a combination of all three, these location-specific fingerprints allow us to reconstruct what the Solar System looked like at its inception. And as we untangle the jumbled mess of the asteroid belt to map the positions of the planetary bodies early in Solar System history, we hope to learn its original structure to better understand its evolution and compare it with the structures we are seeing in other star systems.

Solar System Formation—Were We Pushed?

We know when the Sun formed (~4.568 billion years ago, give or take a few hundred thousand), but what caused that to happen? What started the process of Solar System formation, or more broadly, why are we here? By turning our telescopes to areas of the Milky Way where other stellar systems are in the process of being born, we surmise that our Solar System must have started as a massive cloud of gas and dust that collapsed to form the Sun and planets. But as we know from Sir Isaac Newton and our own repeated experiences with couches and Netflix, something at rest tends to stay at rest. So how, ~4.568 billion years ago, did we go from a massive stagnant floating gas/dust cloud to a rapidly collapsing disk that formed the Sun, planets, asteroids, comets, and eventually, Jar Jar Binks?* Did enough dust grains eventually just randomly bunch together, starting a gravitational domino effect that caused the Sun and planets to form, or was our dust cloud nudged by outside influence? Calling upon outside influence to start the collapse of our solar nebula may sound borderline religious, but the shock wave from a nearby supernova could have been just the nudge needed to transform a diffuse dormant nebula into a rapidly contracting proto-star that would eventually be the Sun. But how can we research something that happened before our Solar System was even born? If the formation of the Solar System began by the shock wave of a supernova, our cloud would also receive other parts of that exploded star, and those other parts could provide physical evidence we can search for in meteorites.

------

* For the record, I am not claiming Jar Jar was a positive outcome, but he was *a* result of the initial collapse of our protostellar cloud . . . and, arguably, the start of the collapse of the Star Wars franchise.

Certain types of meteorites contain what are known as presolar grains, fossils from dead stars that can be found in the most primitive samples. Unfortunately, we do not really know when or where these presolar grains came from, we just know they exist, and they were around prior to the Sun's formation. To find evidence of a particular supernova, one that may have triggered the collapse of our molecular cloud, we would need evidence of when that supernova happened. Lucky for us, supernovae provide us with just the tools to do this in the form of radioactive isotopes that are only created when a star so violently and spectacularly dies.

In cooking, certain dishes require certain conditions. As a native of Kansas City, Missouri, I inherently know you cannot make mouthwatering, fall-off-the-bone BBQ ribs in a microwave: you must cook them slowly in a smoker at low temperature for the better part of a day. Similarly, some isotopes can only be created under certain conditions. And when a star blows up, it creates an environment with a lot of extra neutrons in a confined area. As such, this extremely high density of neutrons creates isotopes that cannot be created any other way. Because some of these isotopes are radioactive, they can act as clocks interred in meteorites, counting down from when that explosion occurred. Evidence of specific radioactive isotopes is precisely the evidence we would need to determine if a supernova were the reason our Solar System started in the first place.

If a star did blow up and seed our molecular cloud with radioactive isotopes, these isotopes would mingle with their cousins of the same element and would therefore be most easily found in phases of meteorites that concentrate that specific element. For instance, if you were looking for evidence of an isotope of iron that was produced only in a supernova, you would want to find phases that had a lot of iron in them to start with. This requires not only finding the right sample and understanding its petrology, but also chemically separating and measuring the isotopic compositions of multiple phases for comparison. Studies to this point remain inconclusive about whether our Solar System was started from the shock wave of a supernova, but stay tuned.

Meteoritics . . . It Takes a Village

The studies outlined above are research projects that have been on-going in the community for years and will continue for years to come. In this sense, these are not necessarily special projects; meteoritic research—and most of science for that matter—is a team sport. This type of research requires networks of experts in diverse fields working in tandem. And this is not just space missions that require thousands of engineers, flight specialists, and mathematicians; this statement is still true for strictly land-based meteorite studies. As an example, take the project of studying whether or not a supernova collapse was responsible for the start of our Solar System. The basic concept is that we want to know how much of a certain type of radioactive material was present at the start of our Solar System. That seems reasonably simple, but first we need to know how much of that radioactive isotope we might have had as a background and how much is produced in different types of supernovae. For that, we need scientists who model these explosions and predict how supernova remnants might mix into a molecular cloud. We don't want to be tricked by alteration in meteorites that may have happened in the last 4.5 billion years, so we need experts in petrology to make sure we are looking at the correct samples. We then need to take the best samples possible to experts in chemistry and mass spectrometry who can isolate and precisely measure the elements and isotopes that are targeted. And this needs to be done multiple times on multiple rocks to verify the results. Incredible progress can be made by one individual on any given topic, but it often takes roomfuls of people with a variety of backgrounds and talents to make meaningful scientific progress on big questions.

The science of meteoritics had a few false starts, and, essentially only sputtered along from its inception over two centuries ago until the Space Age brought the scientific value of meteorites into focus. This abbreviated time span means that only a few generations of specialized scientists have participated in the scientific explosion from studying time capsules of the Solar System. What amazing things will we learn as a larger and more diverse group of researchers begin to study these rocks? Again, stay tuned, or better yet, get involved.

To study meteorites truly is about studying origins—our origins on a habitable planet and our origins as human beings in modern culture. Meteorites have played an outsize role in our journey to humanity and our understanding of how we got here—from the recycled atoms of an exploded star, to the formation of Earth, life crawling out of the ocean, or religious crowds worshipping a new deity—and all of this is meaningful to the project of being human. I am just grateful we are all part of the journey.

Appendices

Notes on the Basics of Meteorite Research

The Taxonomy Man Cometh

As covered in the main chapters, studying the orbital dynamics of asteroids might be able to save our species from an untimely exit, and the scientific information available in meteorites can teach us loads about how we got here in the first place, but that first requires recognition and classification. Meteorites are not just a single type of rock that happens to fall from space, but are, by almost any measure, far more diverse than any rocks formed on Earth. Some meteorites have such high percentages of water that they could have been the source of Earth's oceans, yet some are so dry they gain water just by sitting around on Earth's surface. Other types contain so much organic material that they may have provided the building blocks of Earth's biosphere. Others have never been melted and contain fossils of long-dead stars and the first objects to form in our Solar System. And some look just like rocks erupted from a Hawaiian volcano, but are from Mars. To classify something as a meteorite is just the first step—recognizing that it was not formed on Earth.

Consider watching the original Star Wars movies with the characters having no names, backstories, or connections to any other characters. Sure, Han Solo would still be cool, but he is far more compelling when you find out he made his living as a smuggler, won his ship gambling with a guy named Lando (who later both helps and betrays him), all the while being heavily indebted to a gangster named Jabba the Hut. It is just simply a better story because of the organization and connectivity of the characters. Likewise, the natural world we live in, including the cosmos, is a complicated web of networks. Without organizing and connecting things, we have little hope of understanding big-picture problems or "the story" of our natural world. As such, our best chance of understanding anything in such a complicated system rests on the shoulders of taxonomy, the science of classification.

In terrestrial geology, there are three main classifications into which all rocks fall. Even most nongeologists learned these terms at some point in their science education: igneous, sedimentary, and metamorphic; the tri-pillars of the rock cycle. Of the three, igneous rocks are probably the most exciting to most people, since they derive from volcanoes and, of course, liquid-hot magma. In short, igneous rocks were molten before they crystallized into basically what you see today. They can have lots of shiny crystals and pretty colors, and as such, they get all the ballyhoo and fanfare. Sedimentary rocks are far more unassuming and pedestrian; the common person's rock. They form slowly over time, almost exclusively under water, and they form either from whatever random chunks of stuff that rains down to form layers (think sandstone), or whatever is oversaturated in the water to chemically precipitate (think limestone). Metamorphic rocks are igneous or sedimentary rocks that burned the candle at both ends after they formed. These rocks chose to live a fast and hard lifestyle of pressure and heat, transforming their youthful crystals or layers for twisted and hardened (and sometimes unrecognizable) features only associated with a difficult existence.

The classification scheme described above works well enough when applied to Earth rocks that formed where gravity, running water, and plate tectonics are the primary players that deposit, destroy, and deform them. Yet these familiar conditions of ocean sedimentation and plate tectonics do not exist in space—after all, Earth is the only known planet with oceans of water or plate tectonics. The majority of known meteorites formed in the vacuum of space from coalescing gas and dust and remain largely unchanged, or alternatively, are made of almost solid metal. These are weird rocks. And weird rocks that formed under completely different environmental conditions to Earth rocks require a very different classification scheme.

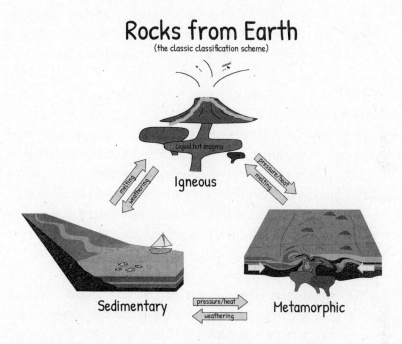

Rocks from Earth
(the classic classification scheme)

This is the basic rock cycle for Earth rocks. Meteorites don't really fit in here. At all.

The well-known science fiction author (and less well-known professor of biochemistry) Isaac Asimov saw taxonomy this way:

The card-player begins by arranging his hand for maximum sense. Scientists do the same with the facts they gather.

Stephen Jay Gould, the influential popular science writer, said:

Classifications are theories about the basis of natural order, not dull catalogues compiled only to avoid chaos.

Yet perhaps my favorite comes from the professor Richard Black-welder,* specialist in tropical beetles and a true giant of modern taxonomy, if there is such a thing.

> *The study of taxonomy in its broadest sense is probably the oldest branch of biology or natural history as well as the basis for all the other branches, since the first step in obtaining any knowledge of things about us is to discriminate between them and to learn to recognize them.*

Classification Is Now in Session

Even early collectors of meteorites knew that rocks from space were very different from Earth rocks. And they knew that if we were to learn much from meteorites at all, it was crucial to recognize the similarities and differences in various samples, allowing us to understand what the various types convey about Solar System formation and evolution. The current classification scheme for meteorites began in the late 1860s with the work of Gustav Rose at the University Museum Berlin and Mervyn Herbert Nevil Story Maskelyne[†] of the British Museum. Perhaps not surprisingly, both of these pioneers were grandsons of famous scientists, and apparently had order and organization in the bloodlines. Gustav was the grandson of Valentine Rose the Elder, who created an alloy of lead, bismuth, and tin that is still known as "Rose metal," though presumably not because of its springtime smell.[‡]

---

* As an aside and to highlight what an interesting (and impressive) taxonomist Black-welder was, after his retirement from his extensive entomological exploits, he chose, in his spare time, to amass everything he could find on the beloved fantasy author J. R. R. Tolkien. Over the next twenty years, Blackwelder personally collected and organized the largest body of Tolkien legendarium ever assembled.

† I would not suggest any, much less all four, of these given names for a modern baby unless you want them mercilessly teased on the school playground and routinely stuffed into various lockers and cabinets.

‡ Maskelyne's grandfather, Rev. Dr. Nevil Maskelyne, was the British Astronomer Royal and was the first person to scientifically measure the weight of the planet Earth.

The Major Classes of Meteorites

In order to have a full appreciation of the kinds of things that we can learn from the many different types of meteorites, it is important to understand how they are broken into their respective groups, and why. Like everything everywhere, there are competing systems. Some people are lumpers, wanting to toss countless meteorites with somewhat similar characteristics together into large bins; some people are splitters, wanting to make hundreds of little groups based on minute differences. Whichever camp you are in, you can choose to group meteorites by physical characteristics, mineralogy, chemical and/or isotopic compositions . . . the list goes on. This list gets even more complicated when you consider the recent advances that have allowed meteorites to be grouped according to "genetic" relationships and where their parent body formed in the disk (close to the Sun versus in the outer Solar System). Things can be a bit confusing in detail, but here we are sticking to a more classical system.

Meteorites can be divided into two main groups, just as Gustav Rose divided them over 150 years ago: chondrites and achondrites. Chondrites got their name because . . . hold on . . . they contain something called chondrules.* Chondrules are not something people, even geologists, outside of the meteorite community are very familiar with, simply because they don't really exist in anything but meteorites. They are complicated in detail but are essentially just millimeter-size round rocks that used to be little molten spheres/globules free-floating in space (hence the round shape). Chondrules make up a large portion of chondrites, often more than 80 percent, and can be readily identified by eye in most chondrites. Achondrites, on the other hand, do not contain chondrules. This is either because they have been melted away and their existence erased, or because they were never there in the first place. Either way, achondrites are pieces of planetary bodies that were molten at

---

* This seemingly primitive prose may give you the impression that space geologists lack creativity, but aside from meteorite groupings, this is not really the case.

some point, forming something roughly planet-like, whose chunks were knocked off and hurled at us.

Classifying Rocks That Never Melted—The Chondrites

Probably the easiest way to think about chondrites of all sorts is that they are sedimentary rocks that formed in space (not underwater, as we are accustomed to thinking with terrestrial sedimentary rocks). Early in the Solar System's history when the Sun was young and planets were only beginning to form, there was a lot of gas, dust, and chunks of small rocks floating around. These seemingly random pieces of celestial sweepings gravitated together, assembling into a relatively small planetoid. Critically, for both the rocks and our classification of them, these chondrite planetoids never produced enough heat to melt their component pieces, whether because their bodies were too small to generate enough heat during their formation,* they formed after most of the short-lived, heat-producing radioactivities present in the early Solar System had largely decayed away,† or both. Regardless of why, we now have unmelted, largely pristine samples of the very early Solar System in the form of chondrites. And because chondrites are largely unaltered since their formation, these samples are critical links to how the Solar System formed and what it was made from prior to secondary processes, like melting, messing things up and making it far more complicated to piece back together.

One of the critical aspects about chondrites in general is that they have largely captured (and maintained) the elemental composition of the molecular cloud from which the Solar System formed. How do we know this? First, we have a good idea of what the Sun is made of thanks

---

* Large bodies form due to the combination of many smaller bodies. When smaller bodies come together in great numbers, there are a lot of collisions that create an immense amount of thermal energy. This conversion from kinetic energy to thermal energy is generally referred to as the "energy of accretion," or "accretionary heat."

† More on this topic later, but it is incredibly cool to think about how isotopes that no longer exist in the Solar System were present when it started. All thanks to the behavior of different types of stars at different points in their life cycles.

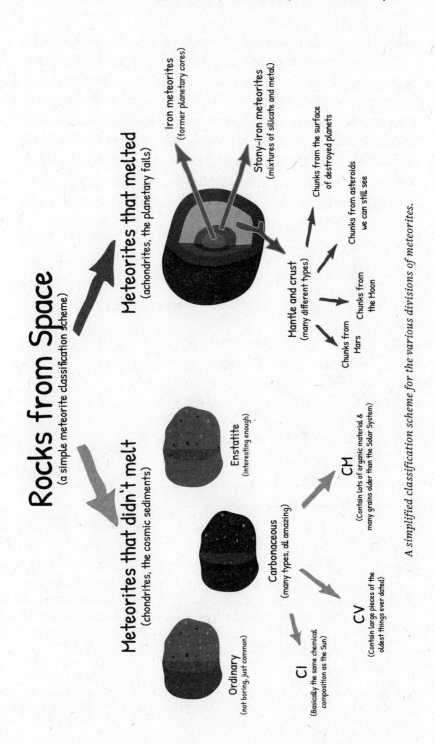

Rocks from Space
(a simple meteorite classification scheme)

Meteorites that melt
(achondrites, the planetary fails)

Iron meteorites
(former planetary cores)

Stony-iron meteorites
(mixtures of silicate and metal)

Mantle and crust
(many different types)

Chunks from the surface
of destroyed planets

Chunks from asteroids
we can still see

Chunks from
the Moon

Chunks from
Mars

Meteorites that didn't melt
(chondrites, the cosmic sediments)

Enstatite
(interesting enough)

Carbonaceous
(many types, all amazing)

Ordinary
(not boring, just common)

CM
(Contain lots of organic material &
many grains older than the Solar System)

CI
(Basically the same chemical
composition as the Sun)

CV
(Contain large pieces of the
oldest things ever dated)

A simplified classification scheme for the various divisions of meteorites.

to a baller technique called spectroscopy. In spectroscopy, basically a machine stares at a star—in this case the Sun—and records the electromagnetic spectrum (that is, visible light, infrared, and ultraviolet radiation) that comes off. From the various wavelengths of this radiation, it is possible to deduce what elements are present, and therefore, what the Sun is made from.* The Sun contains >99.8 percent of the mass of the Solar System, but it is important to know that the remaining scraps that the planets and meteorites were fighting over (Jupiter gobbled up most of it) are still made of the same basic materials and the leftovers

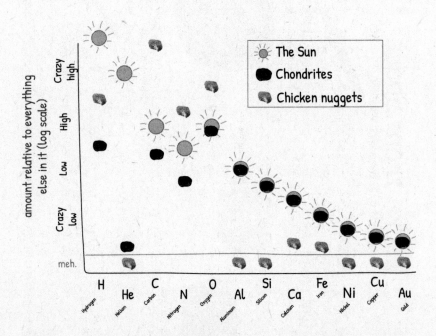

A chart of science proving that chondrites are a much better proxy than chicken nuggets for understanding what the Solar System is made from. The abundance agreement for chondrites and the Sun holds for basically all elements on the periodic table and the only time chondrites significantly differ from the Sun is in highly volatile phases that primarily exist as gases (like helium, for example).

---

* In the Sun's case, it is a whole heck of a lot of hydrogen and helium, some oxygen and carbon, and various trace amounts of the other ~90 naturally occurring elements that combine to form the basis for our existence.

are not some exotic afterthought of Solar System formation. We know this because if we measure what elements the aforementioned chondrites are made from, they have an almost perfect compositional match with the Sun in all but the most volatile elements. And if you think about it, meteorites not containing the same amount of noble gases as the Sun makes sense: elements like helium, neon, and argon carry a full outer electron shell, and so they don't bond into rock-forming minerals.

Basically, the most important thing that this match between chondrite meteorites and the Sun tells us is that the planets and Sun—and everything else in the Solar System—are essentially made from very similar starting materials. It was just the environmental conditions present in each planetary building zone that dictated how much of each element each planet ended up with, and where on each body those elements are now stored. Of course, there were minor transfers of materials that happened later on as material was mixed about, but the gist of it is that we are all made from basically the same stuff.

We have now established that the chondrites are a special breed of primitive rocks, but they are of course not all the same as one another and can be split into many different subgroups. For most, there are three major divisions of chondrites: carbonaceous chondrites, ordinary chondrites, and enstatite chondrites. In some ways, these are thoughtful and informative names that provide immediate information about those classes of rocks. For example, "enstatite" is a well-known mineral that forms when there is not much oxygen around; such meteorites formed under what are known as very "reducing" conditions, or where there was little extra oxygen to spare. In some other ways these main class names are outdated, misleading, and confusing. For instance, the term "carbonaceous" originated because some of the first recognized samples of this group contained abundant carbon, but many of the members of the current group actually contain very little carbon . . . so, not such a good name after all. The dismissive term "ordinary chondrites" slights this class simply because it is the most common type of meteorite to fall to Earth. They are not uninteresting; they are just more abundant than the rest.

The three major divisions of chondrites have much more in common than they have apart. A "typical" chondrite, regardless of whether it is from the carbonaceous, enstatite, or ordinary division, has three major constituents, which you can think of a bit as a timeline of its formation. The oldest constituents of chondrites are the refractory inclusions, which are generally called "CAIs." Whereas this is easily confused with the widely known acronym for the Central Intelligence Agency, "CAI" is actually short for "Calcium-Aluminum-rich Inclusion." This horribly clunky name is at least informative, because, well, as you may have guessed, CAIs are very rich in the elements calcium and aluminum. CAIs are the oldest objects dated in the Solar System and thus their age defines its age. Because CAIs represent time "zero" for the Solar System, they capture a snapshot of what was happening at its birth, so you can think of them as the first picture of our bouncing baby Solar System.*

Chondrules, the rock BBs mentioned earlier as the namesake of chondrites, are the second major phase to emerge during Solar System formation, and therefore the second entry into our Solar System's baby book. Chondrules are a major constituent of chondrites and, more critically, they are an incredibly important phase of Solar System evolution. However, as a community we really don't yet know the exact process(es) that formed chondrules. Some people say chondrules formed from early planetesimal collisions, some say it was from nebular lightning, some say it was from something called "bow shocks" during planet formation. There are many ways researchers will tell you chondrules cannot be made (of course this list includes things like planetesimal collisions, nebular lightning, and bow shocks), but there has yet to emerge a clear mechanism that can satisfy the years of research on the subject. As a running tongue-in-cheek response to when someone asks how chondrules formed, the answer is frequently that "chondrules formed by the

---

* CAIs are what I have spent the bulk of my career to this point studying, so in an active attempt to not get into all the super-interesting specifics, just realize that they are incredibly fascinating objects that give us an immense amount of information about Solar System formation. If you are interested too much beyond that, you will probably need to stray into the scientific literature.

chondrule-forming process."* This now-common response is a nice mix of: (1) obvious truth and humor, (2) recognition of our ignorance on the subject, and (3) embarrassment. But the science marches on, and strides continue to be made on the subject of chondrule formation. After all, we must keep in mind that even with the most generous definition, the field of meteoritics has only existed for a couple hundred years. For comparison, it was many thousands of years before we as humans *really* knew how babies were made, and that is a subject that the vast majority of the adult population appears to very much enjoy doing research on.

The third major phase of a chondrite is the so-called matrix material. No, this matrix is not a series of vertically streaming green numbers in which a beautiful woman in a red dress might appear if you stare at it long enough. The matrix in a meteorite is essentially the "everything else" in a chondrite that is not a CAI or a chondrule. The matrix

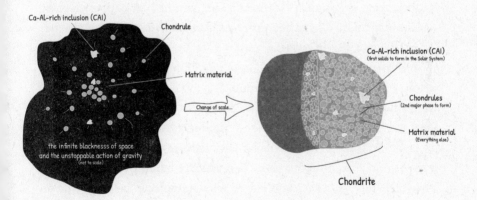

Formation process of a fake chondrite and a cross section showing the three primary phases of a typical fake chondrite.

---

* First quipped by the late John Wasson responding to a heated discussion/question at a Lunar & Planetary Science Conference in the late 1990s.

in chondrites forms at a much lower temperature compared to CAIs or chondrules and as such, contains far more volatile species than its higher temperature counterparts do. In particularly primitive chondrites, the matrix material can host loads of organic compounds; more than eighty types of amino acids have been identified in a single primitive meteorite. The matrix hosts the majority of the water in primitive chondrites, which can be greater than 20 percent of the meteorite's total mass. There is speculation that the water that fills Earth's oceans or the organic material that ignited life and builds Earth's biosphere traces back to matrix material in chondrites. Additionally, another thing found within the matrix in chondrites are something called "presolar grains." These materials exist in different forms in different amounts depending on the type of chondrite, but are in essence fossils from stellar systems that existed before the formation of our Sun. These tiny particles were swept up during the gathering of materials that came together to make our Solar System and everything in it, but escaped melting and incorporation into larger objects like planets, achondrites, or even millimeter-size chondrules. Mixed and melted together, these presolar materials are the building blocks of everything millimeter size and larger in our Solar System. Unmelted, they are our link to what came before it.

All three classes of our chondrule-containing friends (ordinary, enstatite, and carbonaceous) contain the three above-described CAIs, chondrules, and matrix, yet in different proportions and with differing characteristics. This is both due to differences in their original building materials, and to the particular environment each chondrite experienced during its 4.5 billion+ year existence. The individual environments that each chondrite has experienced allow samples to be broken down into types, subgroups, and clans and they are graded based on a combination of the amount of aqueous alteration (or change caused by water, petrologic type 1–2), or heat they have seen (petrologic type 3–6). For example, something deemed a 2 (example, CM2, a carbonaceous meteorite of "M" type) is a carbonaceous meteorite that has seen a moderate amount of water but has not seen high temperatures at any point in its existence. On the other side of things, a type 6 is a sample that has seen lots of heat (up to ≈900°C), which could have come from a number of sources, including internal heat from radiogenic decay, or from impact on the parent body.

The amount of aqueous alteration in a meteorite is important because this describes how much water the meteorite has encountered. At first, this may not seem like it would be important; Earth rocks are exposed to water all the time and are used to it, but for a meteorite, even slightly soppy conditions are a big deal. The major factor in how much water a meteorite has seen is location of formation, essentially how close to the Sun was it when it formed. Meteorites that formed in the inner Solar System (defined as the Sun out to somewhere around the current orbit of Mars) are fairly dry, containing very limited water because the hot Sun drove it all away. However, many meteorites formed beyond what is termed the "snow line," far enough away from the Sun where water ice was abundant and incorporated into any budding planetesimal. Meteorites of petrologic type 1 formed under such conditions and contain around 15 percent water. This is a lot of water, akin to what you see in a used dishrag that was poorly wrung out. These types of me-

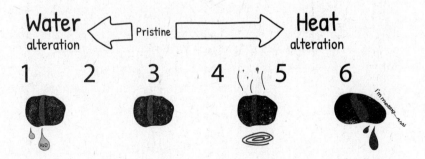

The grading scheme of meteorites.

teorites look and feel wet because they really are.* With this amount of water to start with and >4.5 billion years to react, the effects are noticeable: the boundaries between chondrules are no longer discernible in type 1 chondrites and what remains is a carbon-rich, crumbly pile of what appears to be semi-wet rock flour. As you move up the number scale to the higher petrologic types, the less water (and less aqueous alteration) you have. For instance, in petrologic type 3, you have <3 percent water and these samples maintain chondrules with well-defined borders and clear, unaltered glass present.

If we go beyond petrologic type 3, we are starting to get into rocks that have seen higher temperatures than desired if you hope to maintain a pristine mineralogy. Whereas determining the peak temperature a rock has seen may seem at first a bit arbitrary, it is actually quite easy to understand if you love homemade trail mix as much as I do. If you are smart, you put in chunks of chocolate to complement the peanuts, almonds, raisins, etc. . . . And if you are hiking on a warm day and manage to not eat the entire bag on the way to the trailhead, you will experience what I am talking about. If you reach a certain temperature, the chocolate will just be a bit soft and kind of stick to your hands when you dig in. If the bag gets a bit hotter, the chocolate chunks will have lost their original shape and it is really starting to get a bit messy (I better eat this stuff faster!). If you forgo early snacking, or your hiking buddy is carrying the bag of goodies to keep you from eating it all, your midday hike could result in completely liquefied chocolate coating every raisin and peanut in the bag. This is not altogether a bad thing, as it is still delicious, but does re-inform you that you should carry your own snacks or things like this can happen.

This choco-example is similar to how chondrites are graded for heat exposure. If a meteorite forms at low temperatures, say <100°C, certain mineral structures and phases will be present that indicate lower

---

* It is a common thought that meteorites are flamingly hot all the way through when they land on Earth due to the friction entering the atmosphere. While the outer portion of meteorites do experience a great deal of heat, the interiors are basically unaffected by entry. When meteorites land, they are not usually a hot hunk of rock, but actually very cold, because they have been stored for millions of years at –270 degrees Celsius.

temperatures. As temperatures go up, different mineral structures and phases emerge, oftentimes erasing the lower temperature structures that are no longer stable. Because volatile components such as water can be driven out of the mineral structure as temperatures increase, these events often are nonreversible reactions that can be used as a "high-water mark" for temperatures. As chondrites grade upward from type 3 to type 6, volatile components decrease: water goes from around 3 percent to <1 percent, and there are numerous accompanying mineralogical changes. Melting commences at around 950°C, representing an upper limit of the stability of many minerals: if a chondrite crosses this rock-Rubicon, it will lose its identity and its vaulted status as chondrite, moving itself into the melted world of the achondrites.

Interesting Nuggets About Each Type of Chondrite

Ordinary Chondrites

More than 80 percent of the meteorites found in museums and collections around the globe are classified as ordinary chondrites, so it is easy to see why they are called "ordinary." It remains unclear as to why Earth receives such a high percentage of these samples. It has been logically speculated that the various source asteroids of these meteorites are located at a gravitationally advantageous position in the asteroid belt to perpetually pelt us with whatever gets knocked off them during collision, filling our museums (and eBay) with their planetary flotsam. If such an idea is correct, these parent bodies must be very dynamic and interesting places, because ordinary chondrites have a wide variety of characteristics. Ordinary chondrites range from petrologic types 3–6, showing that some have experienced virtually no thermal alteration (type 3), yet some have seen extensive heating events (type 6). Hydrous alteration has also been identified in some of the least heated ordinary chondrite samples, showing that water has played a minor role for at least some members of this group. Ordinary chondrites also vary widely in the amount of metal present, but even the most metal-poor samples still have visible chunks scattered within them. If you happen to be the type that likes walking around sticking a magnet to everything when you go for a hike, the high metal content of ordinary chondrites makes them comparatively easy to find and recognize as meteoritic, also contributing to their outsize abundance in our collections.

Enstatite Chondrites

Opposite of ordinary chondrites, enstatite chondrites are some of the rarest types of meteorites known, with only ~200 current samples identified. As alluded to earlier, enstatite chondrites have a unique mineralogy that tells us something about their formation region. Because they are among the most reduced* rocks known to exist (containing up to ~10 percent native metal grains), it has been speculated that they formed near the center of the Solar System, where oxygen was very hard to come by, possibly even as central as within the current orbit of Mercury. Another interesting thing about enstatite chondrites is that they are thirsty, having the lowest water content for any rocks studied. Of course, this incredibly low water content also translates into another very sound argument for the idea that they formed close to the Sun.

Carbonaceous Chondrites

I saved the best chondrite class for last. Carbonaceous chondrites are a fairly rare group (<5 percent of total known meteorites) that still manage to find themselves at the center of an overwhelming number of meteoritic studies. There are numerous reasons for the intense focus from the scientific community on carbonaceous chondrites, but it principally boils down to the fact that they are the most primitive of an already primitive class of samples. Simply put, they are archives of unsullied information from the very start of the Solar System. If you have ever been to Phoenix, Arizona, in July, you have likely experienced temperatures higher than some members of the carbonaceous chondrites. Because they are so pristine, these meteorites can provide a view of what our purest building blocks are. Carbonaceous chondrites are the most

---

* Extremely low "free" oxygen around to be used to make oxide minerals.

important hosts of presolar materials, fossils from long-dead stars, allowing scientists to look beyond even the immense 4.567-billion-year history of our own stellar system. Some carbonaceous chondrites contain large and abundant refractory inclusions, CAIs, providing ample material to study that first snapshot of Solar System formation. Additionally, carbonaceous chondrites house an almost inconceivable variety and amount of organic molecules and amino acids, plus a lot of water. This has led many scientists to speculate that the origin of life on Earth is directly tied to the early arrival of such meteorites on our planet. The level of importance carbonaceous chondrites have played on Earth can be reasonably argued, but they are unquestionably important and they are incredibly interesting. From a logistical standpoint, it certainly does not hurt that two of the most important of these types of meteorites are large falls witnessed by humans, both occurring in 1969. In February, >2 tons of the carbonaceous chondrite Allende fell in northern Mexico. And then, to round out the summer of '69, the Murchison meteorite fell in southern Australia in late September. As discussed previously, the falls of these two spectacular meteorites coincided with a world that was abuzz about space science, thanks to the palpable and understandable excitement of landing on the Moon. Laboratories around the world were being built and readied for the return and distribution of lunar samples. People were looking up and excited about outer space, and just at the right time, two incredibly scientifically interesting meteorites arrived in large quantities. All meteorites are special and of course deserve blue ribbons, but from a value standpoint and the information contained within them, carbonaceous chondrites are the bee's knees.

Classifying Rocks That Melted—The Achondrites

As you may have just learned above, achondrites are meteorites that have been exposed to enough heat to melt at some point in their existence, so they do not preserve pristine textures like chondrites often do. In fact, many achondrites are basalts, which are one of the most common types of igneous rocks one might stumble across on tramping over Earth's crust. This fact makes recognizing their meteoritic origin sometimes difficult. But hark! There is an easy way to tell a meteoritic basalt from a terrestrial one: terrestrial basalts do not, as a rule, fall from the sky, whereas meteoritic ones exclusively do. When this method fails and achondritic candidates are located that were not witnessed falls, they undergo a variety of tests to unveil their provenance.[*]

Achondrites may have had their original textures erased by melting, but that does not mean they are all the same. By many measures, achondrites are significantly more diverse compared to their chondritic cousins, representing countless planetary bodies that largely failed to make it all the way to planethood. Those tossed within the fairly broadly grouped term of "achondrites" include chunks of disrupted planets and

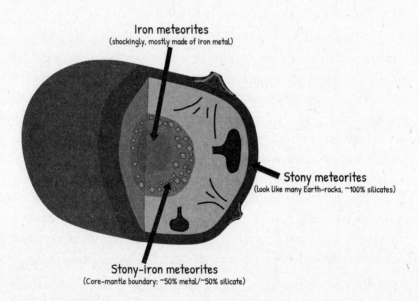

Iron meteorites
(shockingly, mostly made of iron metal)

Stony meteorites
(look like many Earth-rocks, ~100% silicates)

Stony-iron meteorites
(Core-mantle boundary: ~50% metal/~50% silicate)

Cross section of a parent body and the locations where the various types of achondrites derive. Igneous processes (melting) largely create the characteristics of achondrites.

---

[*] Among other techniques, this is done by looking at the iron/nickel ratio of the rock. In general, meteorites have far higher amounts of nickel compared to terrestrial rocks, and this can be a good indicator of its origin.

planetesimals that range from crusts to mantles to iron cores. This list includes the hundreds of pieces of Mars and the Moon (each with their own clear differences) we have in our collections.

Somewhat similarly to chondrites, each achondrite can also be classified by how weathered or fresh it is, or how disturbed by shock it has been (these things were launched off a fairly large chunk of rock, after all). As you might expect, those with low weathering and shock grades are more prized if one is looking for original information about the parent body of the meteorite. However, when layered with other information like the age of the rock, things like weathering and shock can also tell us a great deal about the conditions present over the course of the Solar System. Thus, all space rocks can be informative archives; it just depends on what questions you are most interested in answering and interrogate the rocks appropriately.

Interesting Nuggets About Some Achondrites

Iron Meteorites (Samples of Cores)

When many people think of meteorites, iron meteorites come to mind because meteorite hunters can easily identify them when searching with a metal detector or magnet. As pieces of almost pure metal, iron meteorites are straightforward to identify and recognize as "not normal" rocks. There are some man-made samples that can look similar to iron meteorites; industrial slag waste from steel production, for example, but for the most part, irregular, solid chunks of free-form metal are not the norm in our world and stand out as something special when they are seen on the ground.

Iron meteorites are interesting for several reasons, but first and foremost, they represent something that we have no access to on Earth: planetary core material. The core of Earth, the interior metal marvel that gives us a magnetic field, starts approximately 2,900 kilometers (1,800 miles) below Earth's surface. Since there is 2,900 kilometers of rock with ever-increasing pressure and temperature in the way, you can see this is not an easy thing for us to sample, no matter how interested and motivated one might be. For some perspective on how far off we are from accessing even the outer surface of the outer core of Earth, the deepest hole ever dug—the Kola Superdeep Borehole—is only about 12 kilometers deep. That is less than 0.5 percent of the way there and that project took more than twenty years.* So, as you can see, iron meteorites represent something we do not have access to on Earth, allowing us to much better understand what happens at the center of a planetary body.

Compared to more silicate-rich rocks, iron meteorites are heavily represented in the world's meteorite collections, likely well beyond their relative abundances in space. This is for a couple of reasons. First, they are far more durable than most other rocks and they routinely survive the journey through the atmosphere and earthly impact with the expected grace and robustness you would expect of a lump of iron. Second, as discussed, they are unusual hunks of metal so they are easily recognized and collected as extraterrestrial visitors, whereas silicate-rich samples can better camouflage their origin to collectors.

Stony-Iron Meteorites (Core-Mantle Boundary Region)

I will be honest, I don't really understand stony-iron meteorites, but I promise you, my ignorance is not lonely on this subject. Stony-irons bridge the gap between the iron meteorites of planetary cores and silica-rich, crustal-style rocks we are more familiar with. It is speculated they come from something like the core-mantle boundary layer of destroyed planetesimals, but this is still a hotly debated topic. Because most stony-irons contain approximately equal parts metal and silicate, it makes perfect sense they are placed between the iron meteorites (almost all iron) and the stony meteorites (almost all silicate). A class of stony-irons called "pallasites" are the most famous members of the type and are generally about 50 percent iron-nickel metal, and

---

* The Kola giant hole-a was abandoned due to a combination of higher-than-expected temperatures in the drill hole and the collapse of the Soviet Union, which did not help the funding situation for digging a really deep hole. In case you were looking for a fun vacation destination with the kids, the site is now abandoned *and* an environmental hazard, so it likely will not be crowded.

The beauty that is the Esquel pallasite meteorite.

Esquel – pallasite
Found in Argentina in 1951

about 50 percent of the mineral olivine. This cosmic combo can make pallasites very beautiful to look at: brilliant green olivine crystals encased in a shiny metal matrix make for an otherworldly appearance and command a high price from collectors. Interestingly, it was the unique appearance (and characteristics) of the first known pallasite[*] that helped convince the scientific community that rocks could come from somewhere other than Earth. As you can imagine, this realization was a big deal for the meteoritic community. Other stony-irons like the mesosiderites are less well known, possibly because they are much uglier, and humans are incredibly shallow beings, especially when it comes to rock collecting.

---

[*] This would be the Krasnojarsk meteorite, discovered by the German naturalist Peter Pallas, whom we discussed previously, and the class "pallasite" is named.

Stony Meteorites (Crust and Mantle Samples)

Meteorites that derive from the crust and mantle of their parent bodies are certainly the most difficult to recognize as meteorites. In most cases, they don't look that different from terrestrial crustal samples, and there are undoubtedly untold thousands just lying around waiting to be identified as meteorites. Those that we have recognized as extraterrestrial have truly been a boon for our understanding of all differentiated* bodies in the Solar System. There are many types of achondrites, with the "angrite" group hosting the oldest known igneous rocks in the Solar System, and the "eucrite" group *likely* hailing from the asteroid 4-Vesta, the second-largest object in the asteroid belt. And whereas these and other groups have provided a wealth of scientific information, by far the most sought-after achondrites are meteorites from the Moon and even more so, from Mars.

Similar to Earth, the other large bodies of the Solar System have been continuously pelted by impacts for the last 4.5 billion+ years. Sometimes those impacts are large enough to excavate parts of the surface material where it lands and send various bits hurtling out into space. And, as you may guess, sometimes Earth crosses paths with such liberated planetary material, increasing its ever-expanding alien rock budget to include pieces of the Moon and Mars. If you think it is pretty awesome that we have chunks of rock that were blasted off the surfaces of these bodies, you are right. It is awesome. Of course, for the Moon, we have been there and returned select samples to Earth, which have given us incredible insight into not only how and when the Moon formed, but answers to a bundle of important scientific questions we would have struggled to answer any other way.† Lunar meteorites have greatly complemented this archive of lunar material, and we have >450 identified meteorites that came from the surface of the Moon.

As for Mars, well, Martian meteorites are as of yet the only physical

---

* The term "differentiated" is used to describe a body that has separated itself into crust/mantle and a metallic core.

† These returned lunar samples also showed us that some of the meteorites in our collections were of lunar origin.

pieces of the Red Planet we have to understand its history, so they represent an invaluable resource. As discussed in Chapter 6, based on what we have learned from the multiple unpiloted missions/probes sent to Mars over the decades, Mars has an incredibly captivating history that includes volcanoes, glaciers, periods of liquid water, and environments potentially suitable for the development of life—even prior to when such conditions were present on Earth.

Regardless of the classification any particular space rock belongs to, it is carrying a wealth of information about the formation and/or evolution of our Solar System. Lunar meteorites sample parts of the Moon we have not been to. Pieces of Mars teach us about the long and interesting history of our planetary neighbor. Iron meteorites instruct us on the behavior of planetary cores and inform us about what is, and what happens, far underneath our own feet on Earth. Various types of achondrites illuminate the stories of planets that never were, broken up in violent collisions. Different still are the primitive meteorites that act as snapshots of the earliest epoch of our Solar System, sampling an only slightly modified version of the parent molecular cloud from which the Solar System formed. Indeed, the differences between meteorites are considerable, yet since each unique type teaches us something different about our Solar System history and evolution, they all provide important information we would struggle to gather without their existence.

Appendix 2

Equipment Revolutions

When a new meteorite lands on Earth and pieces make it into the laboratory for study, what happens then? What do researchers do, other than put on cool white outfits and colorful latex gloves? How does one figure out how old a sample is, where it came from, or that it contains amino acids from space? Tons of questions can be asked about any meteorite sample, and modern meteorite researchers have access to a litany of incredible technologies that aid them in this type of research . . . but what types of machines are used, and how the heck do those machines work?

As humans rounded the turn into the nineteenth century, meteorites were starting to be recognized as scientific objects, but the instrumentation and techniques for studying these newly acknowledged items of intrigue were extremely limited by today's standards. For example, in the early 1800s, the principal instrument used to examine meteorites was the human eye,* and the most common scientific procedures done on these objects were measuring their mass and density. These observations and measurements were eventually enough to make the case

---

* I do not mean to discount the impressiveness of the human eye; this was used to great effect to make simple (yet still incredibly important) observations.

that the rocks falling from the sky were not from Earth, but these were incredibly modest scientific techniques. Even at the time, interested researchers knew that the existing technology and methods would need to improve before these enigmatic rocks could be understood at a more advanced level. And while scientific interest is often the engine that drives new technologies in a variety of fields, this narrative is especially evident in the study of meteorites.

For centuries, several themes have permeated the physical science community. And while the most persistent of these is the propensity for its scholars to wear socks with sandals, an almost equally central theme in this field is that its fashion-forward researchers want to find out the contents of what they are studying. It may seem obvious to want to know what something looks like up close or to know the elemental makeup of what you are investigating, but that has not always been possible—and it is not just a coincidence this revolution began around the time meteoritics was becoming a field of study. One of the first major leaps in the study of meteorites—and chemical analysis of samples in general—was something we touched on in Chapter 4. A core piece of evidence that moved the needle toward people realizing these were rocks from space was the work of Edward Howard, who determined that the elements present in the rocks falling from the sky were different from the elements in rocks not falling from the sky. The importance of such analyses on determining the origin of falling stones is hard to overstate, and it was keenly described in 1802 by the French chemist Louis-Nicholas Vauquelin, remarking on Howard's work. And if a French chemist was heaping praise on an Englishman, especially in the early 1800s, it was a pretty big deal:[*]

While all Europe resounded with the reports of stones fallen from the

---

[*] The English and the French have long been adversarial, but keep in mind that their relationship in the eighteenth and nineteenth centuries was dominated by the Anglo-French Wars and just prior to the start of the Napoleonic Wars, both of which primarily pitted the French against the English. So, it was not common practice to toss around kind words about anyone from the other side.

sky, while savants, divided in opinion on this subject, were forming hypotheses to explain the origin of them, each according to his own viewpoint, Mr. Edward Howard, an able English chemist, was pursuing in silence the only route which could lead to a solution of the problems.

What Edward Howard did was advance a technique developed only a few years earlier that enabled scientists to determine the amount of the element nickel present in a sample. This may sound trivial now, but his improvement allowed him to see a chemical distinction between falling rocks (high nickel) and Earth rocks (low nickel). But perhaps even more importantly than showing this distinction, Howard's work showed the potential of chemical investigations to understand a sample's provenance, paving the way for future—and truly remarkable—advancements in chemical analysis yet to come.

In addition to understanding the chemical makeup of a sample, viewing samples up close—much closer than what the human eye can perceive—provides loads of context clues and an understanding of the sample truly on a different level. The next major leap forward for investigating rocks (from space or not) came in the form of something called the "optical" or "petrographic" microscope.

Why not use a normal microscope, you may ask? Normal microscopes pass light through a specimen, such as a thin slice of a leaf or a bug, so the observer can see what is going on at a much higher magnification. Even in the 1800s it was possible to look at individual cells in an organism, helping us understand a multitude of various biological processes. However, just simply magnifying tiny rocks is not super helpful. You might as well just get a bigger chunk of rock and look with the naked eye. You can even cut rocks very thin and look through them at higher magnification like leaves or bugs, but this doesn't have the same benefit for rocks as it does for biologic specimens. Scientists interested in rocks needed to be able to investigate and identify the individual minerals—essentially the cells of rocks—and normal microscopes just did not cut the mustard. The key was light. And not just any light, but something called "polarized light" in particular. Normal, nonpolarized light moves in a wave in three dimensions, having all orientations as it waves along, so when it passes through something like a bamboo leaf, it comes out—not surprisingly—in no particular orientation—just like it went in. This still allows the researcher to enlarge the image through optics and mirrors, but no optical properties are learned using this setup; it is just a very close-up picture of a bamboo leaf.

Similarly, when normal nonpolarized light passes through a thinly cut rock,* it likewise comes out with no particularly diagnostic orientation. The image is enlarged, but that is it. But when polarized light, or light with a unidirectional orientation, is passed through a thinly cut rock, something special happens; each mineral reorients the polarized light in its own unique way, allowing scientists to identify each mineral found in a sample. Of course, minerals reorient nonpolarized light as well, but if the light is oriented in all directions to start with, changing it to a different orientation is not recognized. The "input light" must be standardized, which is what polarization does: it provides a known input. This trick using polarization of light was impressively solved by Scottish physicist William Nicol when, in 1828, he figured out a way to cut a crystal of the mineral calcite in a particular orientation so it produced the polarized light he desired—and provided a path to learning about the optical properties of a rock. When that polarized light was used in conjunction with popular microscopes of the day, bingo-blango, the new science of petrography was invented.

No matter how boring petrography, or "the description and classification of rocks," may sound upon first hearing it, we are all far better off because it exists, and this is acutely true for meteoriticists. Prior to Nicol's invention, about the best we could do was to look really closely at a hand specimen of a rock. Following the invention of the petrographic microscope, it was possible to easily determine at the microscopic scale which type of mineral was where, teaching us how different phases re-

---

* "Thin sections," as they are commonly known, are about 30 microns, or 0.03 millimeters, thick. This is thin enough for light to easily transmit through most minerals, but thick enough to still be reasonably workable.

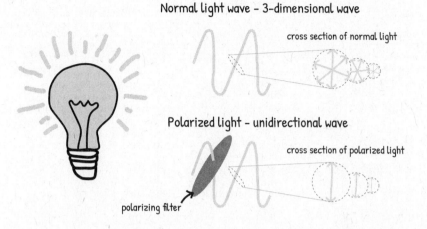

The polarization of light: the key to looking at rocks under the microscope. When polarized light passes through a thinly sliced rock, light reacts differently as it passes through different types of minerals in the rock, giving each mineral identifiable characteristics. Science!

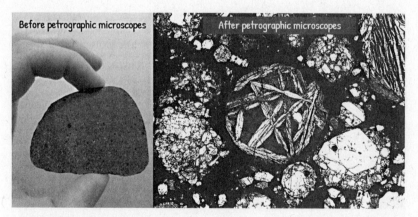

Left: *A chondritic meteorite as it could be seen before the optical microscope.* Right: *A chondritic meteorite magnified in polarized light with visible and identifiable crystals, made possible by the invention of the petrographic microscope. Things got a bit easier for researchers with this new tool.*

acted with one another and which minerals were present (or absent) in different kinds of rocks. The treasure trove of information unleashed by this new microscope was a boon to all sorts of rock hounds, such as volcanologists, sedimentologists, mining prospectors, and, of course, the space rock community.

The twin leaps of improved optics and chemical analysis were the two most important technological advancements that occurred early in the study of meteorites. And while it is not my goal to provide a blow-by-blow history of the steps and leaps that were made in the technology to study rocks, here is a cursory list of the types of instruments and equipment that are commonly used in today's study of space geology.

In the modern study of meteoritics, the two aforementioned branches of optical and chemical analysis still exist, but they have grown much closer together. Certain instrumentation is designed to provide incredibly high-quality optics, and some instrumentation is specially designed to determine the composition of the sample to an almost unimaginable precision. And some instruments can do both at impressive levels. But even though studies of what a sample looks like up close and what a sample is made of can be combined in many cases, divisions still exist regarding the state-of-the-art for each type of information.

First, there are techniques that give information by looking at intact pieces of the rock (commonly called *in situ* techniques, because anything sounds more academic when you throw in a bit of Latin). The huge advantage of *in situ* techniques is that you maintain spatial information about the sample in all its original glory. Researchers can take a sample that has been minimally prepared, view it under magnification, and study how all the different parts are arranged, all the while learning about the basic compositions of all the different pieces that make up the rock. Since there is a ton of information that you can get just from looking at the arrangements and basic compositions of the minerals present in a sample, *in situ* techniques are a critical tool for understanding any sample. Researchers preferring these techniques would argue that the *Mona Lisa* would be a far less impressive painting if art historians put it in a blender to find out exactly what type of paint Leonardo da Vinci used.

But this analogy takes us to the second technique, generally known as "bulk" analysis, where the sample is entirely crushed up and dissolved in acid before it is analyzed so that scientists can find out what it is made of. And while this may sound unnecessarily destructive given the existence of *in situ* techniques, the amount and precision of the data between the two types of analyses can differ by orders of magnitude. Also, depending on the information desired, you don't need to dissolve *that* much material for the bulk technique, so in most cases you are not consuming more than a couple of milligrams of sample—an amount that is barely visible to the human eye. Of course, both *in situ* and bulk techniques have their advantages and disadvantages; choosing the best method of analysis simply depends on what questions you are asking.

Modern methods of chemical analysis are so incredibly easy and efficient compared to what was available even twenty-five years ago, much less 225 years ago, it is uncomfortable to even make comparisons. The pioneering work of chemical analysis done by Edward Howard and many unsung others involved weeks to months of chemistry and preparation in order to learn approximately how much of only a couple major elements were present in a sample. Now it is possible to analyze more than half of the periodic table to better than 5 percent precision in just a few minutes. And, as a nice bonus, in the process you may even get a better than 100,000× magnified picture of what the rock looks like. How do these modern marvels of quantification work? Magic, of course. Well, almost.

When it comes to optical and compositional analyses, it is important to keep in mind that modern instruments vary considerably, as specific instruments are tailored to measure very specific things. Going through all the variations and reasons for them would take books upon books, but let's just stick to the very basics of the machines that are used most routinely in geology, with a focus on meteoritics.

As with a lot of things, the cutting edge of instrumentation technology resides in specialization. While some instruments can do both optics and composition, the absolute best optics machines focus on op-

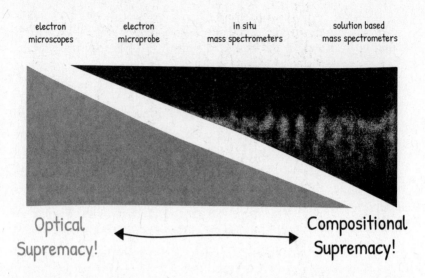

electron microscopes   electron microprobe   in situ mass spectrometers   solution based mass spectrometers

Optical Supremacy! ←————————————→ Compositional Supremacy!

The "best" instrument just depends on what questions you have about the sample. If you want amazing visuals, you won't get the state-of-the-art compositional information. If you want to know the content as well as possible, you will not get visuals. Starting from left to right is an electron microscope, then an electron microprobe, then a mass spectrometer that allows you to look at the sample while measuring it, and finally a mass spectrometer that measures its samples dissolved in acid.

tics, and the absolute best chemical analysis machines focus on chemical analysis.

Starting with the most optically awesome, the electron microscopes are the champs. The scanning electron microscope (SEM) and transmission electron microscope (TEM) are two tools that use—you *guessed it*—electrons to create incredibly magnified images of your sample. While both instruments use beams of electrons to create an image at very high magnification, there are some important differences. First, an SEM emits and bounces electrons off the sample and then collects them on a special coil that allows it to reconstruct the image, providing surface structure information like topographical features. Think of how much you can learn about a planet's structure—mountains and valleys, craters and erosion features—by studying a topographical map of that planet. Now then think of doing it on a spore 1,000 times smaller than you can see. Additionally, because heavy elements like platinum absorb different amounts of electrons than light elements, like silicon, an SEM can also provide some basic compositional information.

Instead of bouncing electrons off the surface to produce an image, the second type of super-microscope, the TEM, transmits electrons through the sample, which is probably why it is so cleverly called the transmission electron microscope. Passing electrons through a sample allows researchers to gain information about its inner structure down to the atomic level. This can inform the researcher about tiny features in the crystalline lattice, providing important information about processes that have acted on the sample, or the existence of hidden nanoscale impurities.

Both types of electron microscopes can produce images about a million times smaller than what the human eye can see. To put that into perspective, if the human eye could see something the width of the state of Colorado, then an electron microscope could see a single bar stool inside the Coors brewery tasting room.

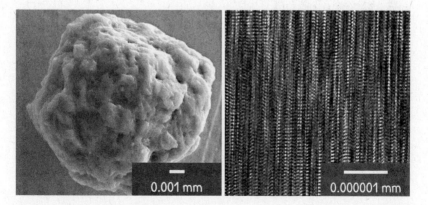

Left: *An SEM image of a presolar silicon carbide grain, a fossil from a long-dead star found in a carbonaceous meteorite.* Right: *A TEM image showing the atomic structure of a mineral called hibonite that formed outside of our Solar System. Note the ridiculousness of the scale bars—these machines are impressive.*

Moving along the line of supercool instruments, we come to something called an electron microprobe. Based on the name, you probably figured out that the microprobe also utilizes the ever-handy electron, but in this case to obtain compositional information instead of ultra-high-resolution imaging. Like the scanning electron microscope, a sample in the chamber of an electron microprobe is also bombarded with electrons. But instead of collecting bounced electrons to make a topographic map, the microprobe is looking for compositional information. And here is where physics really comes in, so bear with me.

All objects, including space rocks, have electrons associated with each atom that makes up the object. In normal circumstances, these electrons are cruising around their respective atoms without much hullabaloo in what is known as the "ground state." However, if you get electrons out of their normal routines, it turns out that you can get a significant amount of information from them. When an electron is momentarily excited by something, it can be given so much energy that it jumps out of its ground state. When the electron returns to its ground state where it wants to be, it emits that extra energy it had in the form of an X-ray. Conveniently, each element has a set of characteristic X-rays associated with the energy loss of its excited electrons, making it possible to figure out what elements are present by the energy signatures of the X-rays emitted—and thus the basic idea of an electron microprobe.* In practice, the sample of interest is loaded into the microprobe vacuum chamber, and the machine focuses a beam of its own electrons at a small point, say a 0.005-millimeter square area. This is enough energy to excite the electrons connected to their respective atoms in that area without damaging the sample. When the fleeting moment of excitement is over for the attached electrons and they drop back to their ground state, the characteristic X-ray signature is captured as the machine scans over a sample, enabling a researcher

---

* I like to think of it in the way that parents always seem to be able to identify when their kid starts crying at a playground among the cacophony of other similar sounds. The kid gets excited and releases a diagnostic wavelength, and the parents detect it. This is the basic premise of an electron microprobe.

Identifying the element oxygen with X-rays

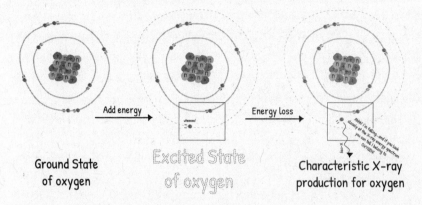

Ground State
of oxygen

Excited State
of oxygen

Characteristic X-ray
production for oxygen

An example of how an X-ray is created and used to identify oxygen in this sample.

to produce a map showing where certain elements are located in the sample.

This incredibly powerful tool works wonders to map the major elements located in a sample, and it is nondestructive, so once the sample is mapped, you can move on and do other sciency things to it, which is a huge plus. The primary downside is that this technique doesn't work on all elements, and the elements generally need to be fairly major constituents of the sample. For instance, if you want to quantify how much silver is in your sample, which is usually well below 0.1 percent of the rock (except in special cases like ore deposits), this method is not going to work very well. However, if you want to find out the quantity of major elements like magnesium, aluminum, calcium, and sodium, or what the major mineralogy phases are in the rock, the electron microprobe is absolutely the best way to go.

Moving farther away from instruments that are built for optical images and toward those built for composition, we start entering a territory where just knowing what elements are present is not quite enough: We now start entering the land of isotopic compositions.

An Aside on Isotopes[*]

If you want a tool that helps you essentially look back in time to trace what happened, isotopes are your meal ticket. This can be something large-scale, like atmospheric changes on Earth, or more local, such as tracking historical changes to a specific lake in upstate New York. If it is a situation where chemical reactions were taking place, isotopes can probably help you figure out what happened. And oftentimes, isotopes can also be used to figure out *when* this event occurred. Even though isotopes are incredibly useful, an unfortunate number of people—geologists included—have nary a clue about what isotopes are or how they can be useful.

An element is determined by how many protons it has in its nucleus. The lovely element oxygen has 8 protons, but it has 3 different versions of itself, as it can have 8, 9, or 10 neutrons, making up the three stable isotopes of oxygen: ^{16}O, ^{17}O, and ^{18}O, where the protons and neutrons are added together to get the mass. These three isotopes, or different versions of oxygen, have the same chemical behavior (because they have 8 protons) but slightly different masses. A cup of coffee can come in multiple sizes, but it still tastes just like coffee. The only real difference is that the smaller size might be easier to carry around. This concept—chemical equivalence in isotopes of the same element, but slight differences in behavior due to their mass—is one of the pillars for why people care about isotopic differences in the first place. Isotopes of an element react *slightly* differently from one another because they have a mass difference. When there is a chemical reaction in nature that involves some of a certain element, the light isotopes of that element will react slightly faster than its heavy isotopes, meaning the light isotopes will be enriched in the product. For instance, clouds are isotopically lighter than the ocean, because the isotopically light versions of oxygen are a little quicker to evaporate and form those clouds.

---

[*] Full disclosure, isotope measurement is my chosen area of study. Again, I will try not to focus too much on this, but precise isotopic measurements have been a transformational tool, particularly in the geosciences. And I am sure I can get at least fifteen to twenty people to agree with me on that.

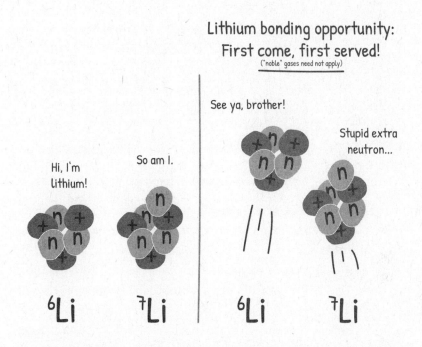

The generalized concept of the difference in isotopes of the same element. Lighter isotopes tend to react faster than their heavier counterparts. This works similarly for all elements that have multiple isotopes, it is just much easier to depict lithium with 3 protons, than say, mercury with 80.

Isotope researchers take advantage of these heavy/light properties and use machines to count the relative amounts of each isotope of a chosen element, and these isotopic differences are records of chemical reactions, whenever they happened. The isotopic differences between samples tell us something about the environments experienced by those samples, sometimes billions of years in the past that we would struggle to learn about any other way.

A second pillar of why isotopic studies are such a big deal is isotopic dating, which is how researchers generally determine the age of a sample. Most people are aware that it is possible to roughly determine the age of a tree by counting its number of rings; and while isotopic dating is a bit more complicated, it again still boils down to counting things, just counting different isotopes. In the case of isotopic dating, there is a radioactive component involved that changes over time, so isotopes related to this radioactive decay are what we are counting. This information is then converted into an age of when something happened, like when the Moon formed, or when a meteorite landed on Earth.

Like an unneutered terrier, the nuclei of certain isotopes naturally have too much energy in them. When a terrier is fixed, they become docile; when unstable isotopes lose that extra energy by radioactive decay, they change into a different element. Thankfully, this all happens predictably, both the resulting element and the rate at which it happens. Perhaps one good (although not 100 percent accurate) way of thinking about this is to compare it with popping popcorn in a pan., If you have 1,000 kernels of popcorn sitting in a hot pan, they are not stable and you know they will eventually pop. You don't know when any single kernel will pop, but you can be pretty sure from the hundreds of times you have popped popcorn in that pan that it will take about two minutes for half of them to be popped. This is the concept of "half-life." If you take that one step further and can stop yourself from eating any of the delicious popped corn, you could, at any point in, say, ~6 minutes from when you started, count each of the popped and each of the nonpopped kernels and, to a reasonable degree of certainty, establish how long the corn had been sitting in the hot pan. This is the concept of isotopic dating. Luckily for people that do isotopic dating, there is not just one type of popcorn, but multiple types that pop at different rates, allowing researchers to compare between various methods.

Counting isotopes is critical if you want to determine the age of anything older than recorded human history, which turns out to be virtually the entire time span of the Solar System. As such, if it were not for the emergence of isotopic dating in the early part of the twentieth century, the age of Earth would still likely be thought to only be a tiny fraction of what it is. Knowing the age of something may seem trivial at first, but consider the effect it has had on other scientific concepts you may be more familiar with. First, in geology, no reasonable person would have ever bought into the idea that ginormous continents such as Africa and South America used to be connected just a few thousand years ago

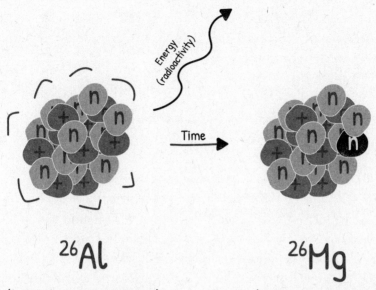

^{26}Al
(too much energy, radioactive)

^{26}Mg
(stable and well adjusted)

Isotopic dating is the backbone of the "how-old-is-this-rock" industry. In this case, the decay of a radioactive isotope of aluminum (13 protons, 13 neutrons, or ^{26}Al) is shown. When ^{26}Al decays, it loses its excess energy, converting a proton into a neutron (shown in dark gray) to become a much more stable isotope of magnesium (12 protons, 14 neutrons, or ^{26}Mg). Since the rate of decay is known, by counting the relative number of ^{26}Mg isotopes in a chosen sample, it is possible to determine how old the rock is.

but are now ~5,000 miles (7,800 km) apart. Those would be very fast-moving continents. However, an Earth over 4 billion years old allows for slow and steady processes to occur, such as ocean basin formation and mountain building, making it possible for a revolutionary and fundamentally important idea like plate tectonics to take hold. Likewise, in biology, there would have been zero chance that the idea of evolution and natural selection would have gained traction had we continued to live on a presumed young Earth. Darwin carefully studied a few groups of finches and noted they developed specialized beaks over numerous generations and hundreds to thousands of years. Imagine the difficulty of convincing even open-minded people that we evolved from a single-celled organism feeding off chemical energy to a fish breathing oxygen to *Homo sapiens* capable of large-scale Nutella production without the benefit of billion-year time scales.

Instruments for Measuring Isotopic Differences

Hopefully the isotopic aside provided good reasons for investigating things beyond incredible visuals and elemental compositions, so now we dive into how isotopes are measured: we use something called a mass spectrometer. There are many types of mass spectrometers, but it is important to have a general concept of what a mass spectrometer does and how it does it before we get into the different types. As the name already implies, this machine produces a "mass spectrum," so that offers a solid start on what it does—a mass spectrometer basically counts isotopes, separating them into the different masses that make up whatever sample we are investigating. So how does it do that? It involves a giant magnet. And it is awesome.

Perhaps the most important term to understand in mass spectrometry is "ion," because almost everything is based on ions. An ion is just an atom that has a positive or negative charge,* but, importantly, once an atom has a charge, it can be moved and steered around using electric and magnetic fields. Although there are big differences between mass spectrometers in how they create ions from the sample they are studying, all mass spectrometers work on the same basic principles: (1) the machine creates ions, (2) ions are steered around a corner where heavy things turn less sharply than light things, (3) the machine counts the number of heavy and light things. This turning concept is not hard to imagine, as it is simply based on momentum. If you have a Saint Bernard and a Chihuahua both running together in a line and throw a steak to one side of them, the small Chihuahua will turn much faster than the

---

* In a panic, Atom 1 says to a friend: "I just lost an electron!"
  Friend: "Are you sure? You look fine to me."
  Atom 1: "I'm positive."

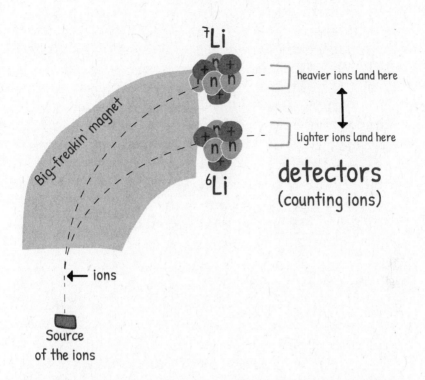

The basic concept of a mass spectrometer: momentum. Heavy things are harder to turn than light things, thus moving these things around a corner will separate them by mass. Modern mass spectrometers can simultaneously measure more than 8–12 masses at the same time, instead of only the two shown here.

more massive Saint Bernard. The Saint Bernard will then eat the steak and Chihuahua both, and this analogy has completely lost its relevance to physics.

Even though isotopic work continues us along the path to compositional supremacy, we do not yet have to abandon sample visuals thanks to an optically and isotopically capable instrument called a "secondary ion mass spectrometer," or SIMS. If you are looking to get both high-quality visuals and good isotopic data from a sample, this is your horse. A SIMS instrument is a combination of a microscope and a mass spectrometer, so you use the microscope to locate specific interesting phases and target those phases for isotopic measurement. Essentially,

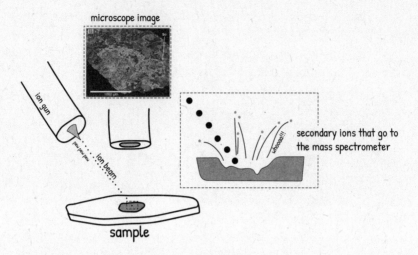

microscope image

ion gun

pew pew pew

ion beam

secondary ions that go to
the mass spectrometer

whooaa!!!

sample

*A secondary ion mass spectrometer (SIMS) instrument in action. First, you find
where to aim using the microscope feature, and second, you blast away tiny bits of
the sample to send them into the mass spectrometer.*

once you find something interesting in the microscope you want to
know the isotopic composition of, you turn on a gun of high-energy ions
(*really* tiny bullets), focus the ions on your target, and blast the part of
the sample you want to measure. The material getting blasted forms its
own charged particles—hence "secondary ion" in the name—and these
secondary ions are then sucked into the mass spectrometer portion of
the instrument, which separates the ions by mass, giving us the information we are looking for.

There are many upsides to SIMS instruments. Most of all, you get
spatial information that is paired with isotopic data, a huge plus. However, the downside to this method is that when you blast away at the
sample, you ionize many more elements than the one you were prepared to look at in the mass spectrometer, and these ionized interlopers can hurt the overall achievable precision. Try throwing a handful of
Skittles in the air and only trying to catch a green one in your mouth. It
would be easier to not taste the entire rainbow at once but to only have
green ones to start with.

This brings us to how to get the highest achievable precision of isotope analysis—we must leave the realm of optical images and move into solution chemistry. Unfortunately, this means physically destroying the sample you are interested in by crushing it and dissolving it in acid; so, if you want to see what your sample looks like under the microscope, you should probably start with that step. If we target an element for isotopic measurement, other elements present get in the way, like too many Skittles in the air. Therefore, we must first chemically isolate the element we are interested in measuring, essentially pouring out the bag of Skittles and picking out only green ones before we toss them skyward. And for geology, that means performing chemistry to separate the element we want from other elements that might interfere with the measurement. This then takes us to the compositional kings, the solution-based mass spectrometers.

Once you have spent weeks to months getting your sample in solution and performing chemistry to isolate your element of interest, you can move on to the solution-based mass spectrometer that *best suits* your needs. The two main types are the thermal ionization mass spectrometer and the plasma ionization mass spectrometer, which work better or worse depending on the specific element you want to measure. The only real difference in the machines is again *how* you ionize. The thermal machine requires you load your sample on a filament (which is very similar to a traditional lightbulb filament). The machine then provides energy to the filament, which ionizes the sample using heat from the now-glowing filament. The second type of machine, the plasma machine, essentially makes a flame using the element argon to produce a tiny area about twice the temperature of the Sun's surface, ionizing your sample as it sucks it up as a solution. Regardless of the introduction method, solution-based mass spectrometers can produce mind-blowing isotopic precision on a huge variety of elements and are a critical tool to help answer questions of great interest to the broader community.

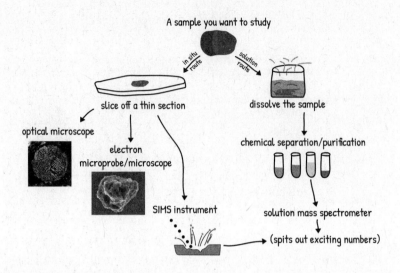

Possible progressions one might take to study your super awesome rock sample using the various instruments described here. It is certainly possible to do multiple paths if you have the opportunity and time.

Another Aside: The Study of Organic Compounds in Meteorites

Unlike elements, which cannot be broken down into simpler forms, organic molecules are collections of elements that are assembled in myriad ways around various numbers of carbon atoms. As such, when studying organic molecules, maintaining the original structure of the molecule is paramount; you can't just dissolve the whole rock in boiling acids or ionize it completely, as this would destroy many of the compounds you are looking to study. Luckily, however, many of the methods discussed above also apply to the study of organic matter in meteorites, albeit with a few gentle twists. Just like for elemental analyses, organic matter research is also divided into *in situ* techniques and bulk techniques, and the same basic benefits and disadvantages apply. You either get really cool visual information about where the organic material is in the sample (*in situ* techniques), or you can precisely quantify the type and amount of organic matter in a sample (bulk techniques).

Meteorites contain two major types of organic compounds: those molecules that dissolve in solvents like water and alcohol, cleverly called soluble organic matter, and those that do not dissolve even after harsh acid treatments, cleverly called insoluble organic matter. Separating insoluble material for study is not that difficult; you just crush the meteorite and expose it to a slurry of acids, and the goo that is leftover when all the rock dissolves is the insoluble matter. Typical burning-down-the-haystack-to-find-the-needle stuff. Getting at the less robust soluble organic matter requires leaching a powdered meteorite sample with a series of solvents to separate—and not destroy—the soluble organic matter. Once the targeted type of organic material is separated from the rest of the meteorite and concentrated, the real fun begins . . . finding out what molecules are present.

Determining the type of organic material in a sample can be done in several ways, but one of the most common is our old friend, mass spectrometry. Mass spectrometry for organic molecules follows the same basic principle as described earlier for elements—you still separate things by mass—but instead of breaking everything up into its individual atoms, you only break certain types of bonds in the organic goo using specific chemicals. You then run the material through a specialized mass spectrometer that not only measures the mass of the broken-up goo chunks but also allows you to calculate how the pieces fit back together. This reconstruction of molecules is possible because you know which bonds were intentionally broken during the process.*

A second common method of determining the type of organic material is what is known as nuclear magnetic resonance, or NMR. This is the same technique used in medical imaging, but since people are afraid of anything that has the word "nuclear" attached to it, the medical field decided to drop that part and just call it magnetic resonance imaging, or

---

* I like to think of this as an uncut puzzle you can't see. However, you can see individual puzzle pieces after you cut it. You figure out how the pieces fit back together because you decided how to cut them up in the first place.

MRI.* This technique leverages the inherent magnetic spin properties of protons and applies a strong magnetic field to excite them. The energy transfer caused by the applied magnetic field on the sample produces measurable signatures that can be used to reconstruct which molecules are present in the sample.

As discussed previously, researchers have used both *in situ* and bulk techniques to recognize and isolate hundreds of organic molecules that are not found in Earth's biosphere, as well as many compounds found in meteorites that are essential to life on Earth. Scientists building machines to figure out how we were built: it doesn't get much cooler than that.

Just based on the technology required for some types of studies, meteorites can be very expensive things to work on: modern machines are hundreds of thousands to multimillion-dollar investments that require specialized lab space, clean-room facilities that keep out contamination, and technical experts to oversee it all. These machines and infrastructure require major investments by national laboratories, universities, and research institutes around the world. But these investments have huge returns. The instruments that have been invented and improved upon over the last ~225 years to study meteorites and other geologic samples have revolutionized how we think about the Earth and Solar System, and, at the same time, have resulted in technological advancements and engineering feats that have benefited countless other industries and research fields. All the while training a continual stream of critical thinkers and problem solvers.

Scientific progress, regardless of the field, takes more than creative ideas, cool machines, and dedicated people; it also requires dedicated funding to make it all happen. Science as a field is amazing, inspiring, exciting, fundamental to development . . . and hundreds of other adjectives, but it is not free. Certainly, the machines and infrastructure discussed above cost money, but so do materials. And even though scientists generally love what they do, they still must eat. So beyond large

---

* Marketing. Really. Important.

capital purchases, where does the money come from to pay for meteorite research?

Whereas there are a few smaller private grants and pots of money here and there around the world, the largest funders of meteorite-related research in the world are the European Research Council, the Japanese space agency JAXA, and, by far the largest single funder, NASA. To put things in monetary perspective, of NASA's ~$20 billion budget, NASA allocates approximately $10 million per year (~0.05 percent of its total) for things related to meteorite research. This is not just a research budget for different groups around the country, but is everything together, including finding and curating meteorites, portions of salaries for professors-postdocs-graduate students, help with the costs of said researchers going to and hosting conferences, costs associated with sample analysis, costs of acquiring samples, publishing results . . . everything. And while $10 million a year may seem like a large sum of money, keep in mind that telling the origin story of X-Men's Wolverine in 2009 cost ~$150 million, and Wolverine is a fictional character in the Marvel Universe. I would argue that knowing our origin story, the Solar System's origin story in *the* Universe, is easily worth the investment.

Without this advanced instrumentation and a continuous stream of monetary support for science, our understanding of the natural world around us would be significantly diminished. And while specialized machines represent some of the modern tools and methods used in the study of meteorites that often result in high-impact research, there is still an incredible amount of critical information gained daily by researchers using relatively simple petrographic microscopes and a keen human eye. In order to make progress in any field, it takes a wide range of people with diverse talents and ideas. The supercool instruments discussed above represent one way of doing cutting-edge meteoritic research, but this field still holds many exciting questions about the Solar System and our origins that remain unanswered.

Acknowledgments

There are a lot of things that I am incredibly grateful for, but since I like to do things chronologically, I will start with my parents. Not because of anything specific, really, but just because I am extremely lucky to have them as parents and because they are simply great people and have been my entire life, at least. Nobody that has met my parents thinks anything else—I am absolutely positive of that—and they make the planet a far better place for everyone just by being here. Next, I want to thank my lovely wife, Celeste. Sure, she has a cool cosmic-sounding name, but she also has been a positive force in every step of this book, not to mention for a good chunk of my adulthood. She always manages to provide solid advice, unrelenting support, and makes me feel happy about life. You are great.

Next, I want to thank Jan Render, Quinn Shollenberger, and Tim and Anne Riley. Somehow, without complaint, you read every mediocre first draft of every chapter and your honest and thoughtful comments made every single chapter better. Also, I would like to thank Rebekah Hines, Emilie Dunham, Sarah Mazza, Zita Martins, and Conel Alexander for giving extremely helpful comments on select chapters.

Additionally, I would like to thank Natalie Starkey and Lewis Dartnell for their early support about my idea for the book and to Lewis for introducing me to Ian Bonaparte, who is an absolute superstar of an agent that made the nebulous process of publishing this book very understandable, enjoyable, and possible. I would also like to thank my editor, Mauro DiPreta, for taking a chance on a first-time wannabe-author, for making helpful comments to greatly improve the book, but also letting me be me in the writing.

Furthermore, I want to acknowledge the kindness of the many people I contacted about specific questions or for use of their photographs or art. You all enhanced this book significantly, and I am incredibly grateful.

And since we rarely get the opportunity to thank our friends and mentors as we meander through our lives and careers, I want to mention just a few of the people who have not already been mentioned above. This crew, and so many others that are not mentioned, have always inspired me to do cool science, be a good person, and genuinely enjoy life—all at the same time: Ariel Anbar, Brad Divelbiss, Erick Ramon, Geoff Brennecka, Ian Hutcheon, Keith Morrison, Lars Borg, Levi Maxwell, Mini Wadhwa, Patrick Noonan, Peter Meserve, Ryan Smith, Sandra Peters, Thomas Kruijer, Thorsten Kleine, Virginia Barrett, and always last alphabetically by first name, Zac Gonsior. Thank you for being you.

Finally, I thank the reader for your interest in science and for taking the time to read about how meteorites have shaped our world. I am fortunate to be part of a small but great community of researchers, and it is a thrill to be able to share some of our contributions with you.

Sources and Additional Reading

Resources I Leaned on Heavily for Research

Bevan, A., and J. De Laeter. (2002). *Meteorites: A Journey Through Space and Time.* University of New South Wales Press, Sydney.

Burke, J. G. (1986). *Cosmic Debris: Meteorites in History.* Berkeley: University of California Press.

Golia, M. (2015). *Meteorite.* London: Reaktion Books.

McCall, G. J. H., A. J. Bowden, and R. J. Howarth (eds.). (2006). "The History of Meteoritics and Key Meteorite Collections: Fireballs, Falls and Finds." Geological Society, London, Special Publications, 256, 305–23.

Nield, T. (2011). *The Falling Sky: The Science and History of Meteorites and Why We Should Learn to Love Them.* Guilford, CT: Lyons Press.

Starkey, N. (2018). *Catching Stardust: Comets, Asteroids and the Birth of the Solar System.* Bloomsbury.

Basically, any article written by Ursula Marvin.

The brilliant Wikipedia (https://www.wikipedia.org/) and wikimedia commons (https://commons.wikimedia.org) for copious information and many pictures.

Meteoritical Bulletin Database (https://www.lpi.usra.edu/meteor/).

Some Specifics Beyond the Above or Those Referenced in the Text

Introduction

Güdel, M. (2007). "The Sun in Time: Activity and Environment." *Living Reviews in Solar Physics* 4, 3.

Information about alligator farming in Florida: https://myfwc.com/wildlifehabitats/wildlife/alligator/farming/.

Chapter 1

Alvarez, L.W., W. Alvarez, F. Asaro, and H. V. Michel. (1980). "Extraterrestrial Cause for the Cretaceous–Tertiary Extinction." *Science* 208 (4448): 1095–1108.

Flanders, S.E. (1962). "Did the Caterpillar Exterminate the Giant Reptile?" *Journal of Research on the Lepidoptera* 1, no. 1: 85–88.

Hildebrand, A. R., G. T. Penfield, et al. (1991). "Chicxulub Crater: A Possible Cretaceous/Tertiary Boundary Impact Crater on the Yucatán Peninsula, Mexico." *Geology* 19, no. 9: 867–71.

Hull, P. M., et al. (2020). "On Impact and Volcanism across the Cretaceous-Paleogene Boundary." *Science* 367, 266–72.

Ivany, Linda C., William P. Patterson, and Kyger C. Lohmann. (2000). "Cooler Winters as a Possible Cause of Mass Extinctions at the Eocene/Oligocene Boundary." *Nature* 407: 887–90.

Levison, H. F., et al. (2009). "Contamination of the Asteroid Belt by Primordial Trans-Neptunian Objects." *Nature* 460: 364–66.

Walsh, K. J., A. Morbidelli, S. N. Raymond, D. P. O'Brien, and A. M. Mandell. (2011). "A Low Mass for Mars from Jupiter's Early Gas-Driven Migration." *Nature* 475: 206–9.

Wielicki, M., M. Harrison, and D. Stockli. (2014). "Popigai Impact and the Eocene/Oligocene Boundary Mass Extinction." Goldschmidt abs. #2704.

Zahnle, K., et al. (2007). "Emergence of a Habitable Planet." *Space Science Review* 129: 35–78.

Zahnle, K., et al. (2010). "Earth's Earliest Atmospheres." *Cold Spring Harbor Perspectives on Biology* 2: a004895.

Chapter 2

Xi Z-z. (1984). "The Cometary Atlas in the Silk Book of the Han Toeb at Mawangui." *Chinese Astronomy and Astrophysics* 8, 1–7.

Chapter 3

Barrows, T. T., et al. (2019). "The Age of Wolfe Creek Meteorite Crater (Kandimalal), Western Australia." *Meteoritics & Planetary Science* 54, 2686–97.

Bevan, A. W. R., and P. Bindon. (1996). "Australian Aborigines and Meteorites." *Records of the Western Australian Museum* 18, 93–101.

Bjorkman J. K. (1973). "Meteors and Meteorites in the Ancient Near East." *Meteoritics* 8, no. 2.

Buchner, E., et al. (2012). "Buddha from Space—An Ancient Object of Art Made of a Chinga Iron Meteorite Fragment." *Meteoritics & Planetary Science* 47, 1491–1501.

Comelli, D., et al. (2016). "The Meteoritic Origin of Tutankhamun's Iron Dagger Blade." *Meteoritics & Planetary Science* 51, no. 7: 1301–9.

D'Orazio, M. (2007). "Meteorite Records in the Ancient Greek and Latin Literature: Between History and Myth." Geological Society, London, Special Publications, 2007, vol. 273, pp. 215–25.

Gettens, R. J., R. S. Clarke Jr., and W. T. Chase. (1971). "Two Early Chinese Bronze Weapons with Meteoritic Iron Blades." *Occasional Papers* 4, 1e77, Freer Gallery of Art Washington, D.C.

Hamacher, D. W. (2014). "Comet and Meteorite Traditions of Aboriginal Australians." *Encyclopaedia of the History of Science, Technology, and Medicine in Non-Western Cultures,* 2014.

Hamacher, D. W., and R. P. Norris. (2009). "Australian Aboriginal Geomythology: Eyewitness Accounts of Cosmic Impacts?" *Archaeoastronomy,* 22, 60–93.

Hartmann, W. K. (2015). "Chelyabinsk, Zond IV, and a Possible First-Century Fireball of Historical Importance." *Meteoritics & Planetary Science* 50, 368–81.

Jambon, A. (2017). "Bronze Age Iron: Meteoritic or Not? A Chemical Strategy." *Journal of Archaeological Science* 88, 47–53.

Johnson, D., et al. (2013). "Analysis of a Prehistoric Egyptian Iron Bead with Implications for the Use and Perception of Meteorite Iron in Ancient Egypt." *Meteoritics & Planetary Science* 48, no. 6, 997–1006.

Kohman, T. P., and P. S. Goel. (1963). "Terrestrial Ages of Meteorites from Cosmogenic C14." In *Radioactive Dating,* International Atomic Energy Agency, Vienna, 395–411.

Marvin, U. (1992). "The Meteorite of Ensisheim: 1492 to 1992." *Meteoritics* 27, 28–72.

Marvin, U. (1996). "Ernst Florens Friedrich Chladni (1756–1827) and the Origins of Modern Meteorite Research." *Meteoritics & Planetary Science* 31, 545–88.

Photos, E. (1989). "The Question of Meteoritic Versus Smelted Nickel-Rich Iron: Archaeological Evidence and Experimental Results." *World Archaeology* 20, no. 3, *Archaeometallurgy* (February 1989), pp. 403–21.

Pillinger, C. T., and J. M. Pillinger. (1996). "The Wold Cottage Meteorite: Not Just Any Ordinary Chondrite." *Meteoritics & Planetary Science* 31, 589–605.

Remler, P. (2010). *Egyptian Mythology A to Z.* 3rd ed. New York: Chelsea House.

Thomsen, E. (1980). "New Light on the Origin of the Holy Black Stone of the Ka'ba." *Meteoritics* 15, 87–91.

Wainwright, G. A. (1932). "Iron in Egypt." *Journal of Egyptian Archaeology* 18, 3–15.

Yau, K., et al. (1994). "Meteorite Falls in China and Some Related Human Casualty Events." *Meteoritics & Planetary Science* 29, 864–71.

Chapter 4

Sears, D. W. (1975). "Sketches in the History of Meteoritics: The Birth of the Science." *Meteoritics* 10, 3, 215–25.

Sears, D. W., and H. Sears. (1977). "Sketches in the History of Meteoritics 2: The Early Chemical and Mineralogical Work." *Meteoritics* 12, 1, 27–46.

Chapter 5

Bland, P. A., et al. (1996). "The Flux of Meteorites to the Earth over the Last 50,000 Years." *Monthly Notices of the Royal Astronomical Society* 283, 551–65.

Chyba, C., and C. Sagan. (1992). "Endogenous Production, Exogenous Delivery and Impact-Shock Synthesis of Organic Molecules: An Inventory for the Origins of Life." *Nature* 355, 125–32.

Dodd, M. S., et al. (2017). "Evidence for Early Life in Earth's Oldest Hydrothermal Vent Precipitates." *Nature* 543, 60–65.

Elsila, J. E., et al. (2016). "Meteoritic Amino Acids: Diversity in Compositions Reflects Parent Body Histories." *ACS Central Science* 2016, 2, 6, 370–79.

Evatt, G. W., et al. (2020). "The Spatial Flux of Earth's Meteorite Falls Found via Antarctic Data." *Geology* 48, G46733.1.

Hashimoto, G. L., et al. (2007). "The Chemical Composition of the Early Terrestrial Atmosphere: Formation of a Reducing Atmosphere From CI-like Material." *Journal of Geophysical Research* 112, E05010.

Iglesias-Groth, S., et al. (2011). "Amino Acids in Comets and Meteorites: Stability under Gamma Radiation and Preservation of the Enantiomeric Excess." *Monthly Notices of the Royal Astronomical Society* 410, 1447–53.

Kitadai, N., and S. Maruyama. (2018). "Origins of Building Blocks of Life: A Review." *Geoscience Frontiers* 9, 1117–53.

Martins, Z. (2019). "Organic Molecules in Meteorites and Their Astrobiological Significance." In *Handbook of Astrobiology* (CRC Press, Boca Raton, 2019), 177–94.

Pizzarello, S., G. Cooper, and G. Flynn. (2006). "The Nature and Distribution of the Organic Material in Carbonaceous Chondrites and Interplanetary Dust Particles." In *Meteorites and the Early Solar System II,* edited by D. Lauretta, L. Leshin, and H. McSween Jr. (Tucson, University of Arizona Press), 625–51.

Pizzarello, S., and E. Shock. (2010). "The Organic Composition of Carbonaceous Meteorites: The Evolutionary Story Ahead of Biochemistry." *Cold Spring Harbor Perspectives in Biology* 2:a002105.

Prasad, M. S., et al. (2013). "Micrometeorite Flux on Earth during the Last ~50,000 Years." *Journal of Geophysical Research: Planets* 118, 2381–99.

Ritson, D. J., et al. (2020). "Supply of Phosphate to Early Earth by

Photogeochemistry after Meteoritic Weathering." *Nature Geoscience* 13, 344–48.

Weiss, I. M., et al. (2018). "Thermal Decomposition of the Amino Acids Glycine, Cysteine, Aspartic Acid, Asparagine, Glutamic Acid, Glutamine, Arginine and Histidine." *BMC Biophysics* 11: 2.

Chapter 6

Borg, L. E., et al. (1996). "The Age of the Carbonates in Martian Meteorite ALH84001." *Science* 286, 90–94.

Brennecka, G. A., et al. (2014). "Insights into the Martian Mantle: The Age and Isotopics of the Meteorite Fall Tissint." *Meteoritics & Planetary Science* 49, 412–18.

Cassata, W. S., et al. (2012). "Trapped Ar Isotopes in Meteorite ALH84001 Indicate Mars Did Not Have a Thick Ancient Atmosphere." *Icarus* 221, 461–65.

Head, J. (2012). "Mars Climate History: A Geological Perspective." Lunar and Planetary Science Conference Abstract #2582.

McKay, D. S., et al. (1996). "Search for Past Life on Mars: Possible Relic Biogenic Activity in Martian Meteorite ALH84001." *Science* 273, 924–30.

McSween, H. Y., Jr. (1994). "What We Have Learned About Mars from SNC Meteorites." *Meteoritics* 29, 757–79.

Merino, N., et al. (2019). "Living at the Extremes: Extremophiles and the Limits of Life in a Planetary Context." *Frontiers in Microbiology*, 10, doi:10.3389/fmicb.2019.00780.

NASA. Mars Meteorite Compendium.

Rampelotto, P. H. (2013). "Extremophiles and Extreme Environments." *Life* 3, 482–85.

Schopf, J. W. (1999) "Life on Mars: Tempest in a Teapot? A First-Hand Account." *Proceedings of the American Philosophical Society* 143, 359–78.

Chapter 7

Drouard, A., et al. (2019). "The Meteorite Flux of the Past 2 M.Y. Recorded in the Atacama Desert." *Geology* 47, 673–76.

https://www.nationalgeographic.com/science/article/130215-meteorite-hunter -russian-moon-interstellar-rocks-space-meteor-asteroid.

Chapter 8

Unsalan, O., et al. (2020). "Earliest Evidence of a Death and Injury by a Meteorite." *Meteoritics & Planetary Science*, 1–9.

Yau, K., et al. (1994). "Meteorite Falls in China and Some Related Human Casualty Events." *Meteoritics & Planetary Science* 29, 864–71.

Chapter 9

Bland, P. A., et al. (2009). "An Anomalous Basaltic Meteorite from the Innermost Main Belt." *Science* 325, 1525–27.

Bryson, J. F. J. et al. (2020). "Constraints on the Distances and Timescales of Solid Migration in the Early Solar System from Meteorite Magnetism." *Astrophysical Journal* 896, 103.

Cameron, A. G. W., and J. W. Truran. (1977). "The Supernova Trigger for Formation of the Solar System." *Icarus* 30, 447–61.

Ceplecha, Z. (1961). "Multiple Fall of Pribram Meteorites Photographed. Double-Station Photographs of the Fireball and Their Relations to the Found Meteorite." *Bulletin of the Astronomical Institute of Czechoslovakia* 12, 21–47.

Gemelli, M., et al. (2014). "Chemical Analysis of Iron Meteorites Using a Hand-Held X-Ray Fluorescence Spectrometer." *Geostandards and Geoanalytical Research* 39, 55–69.

Glavin, D. P., et al. (2018). "The Origin and Evolution of Organic Matter in Carbonaceous Chondrites and Links to Their Parent Bodies." In *Primitive Meteorites and Asteroids*, ed. N. Abreu (Amsterdam: Elsevier), 205–71.

Render, J. H., and G. A. Brennecka. (2021). "Isotopic Signatures as Tools to Reconstruct the Primordial Architecture of the Solar System." *Earth and Planetary Science Letters* 555, 116705.

Appendix 1

Uluozlu, O. D., et al. (2009). "Assessment of Trace Element Contents of Chicken Products from Turkey." *Journal of Hazardous Materials* 163, 982–87.

Appendix 2

Howard, E. (1802). "Experiments and Observations on Certain Stony and Metalline Substances, Which at Different Times Are Said to Have Fallen on the Earth; Also on Various Kinds of Native Iron."

Photo Credits

Illustrations and photographs provided by the author except for those listed below.

Page 25 (top): Steve Jurvetson from Menlo Park, USA, CC BY 2.0

Page 30: Lawrence Berkeley Laboratory

Page 39: Luc Viatour

Page 42: National Portrait Gallery

Page 44: Andrew Harnik/AP

Page 45: NASA, ESA, J. Hester and A. Loll (Arizona State University)

Page 48: Public domain

Page 49: Giotto Di Bondone, public domain

Page 59 (top): © Griffith Institute, University of Oxford

Page 59 (bottom): Mark Fischer, CC BY-SA 2.0

Page 66: Stipich Béla, CC BY-SA 3.0

Page 68: Freer Gallery of Art, Smithsonian Institution, Washington, D.C.: Purchase—Charles Lang Freer Endowment, F1934.10 and F1934.11a-c.

Page 72 (top): Dainis Dravins—Lund Observatory, Sweden

Page 72 (bottom): By Boxer Milner, a Djaru Elder from Billiluna, WA. [https://web.sas.upenn.edu/psanday/exhibition/painting-gallery/]

Page 74: "Persia by a Persian: being personal experiences, manners, customs, habits, religious and social life in Persia." Author: Isaac Adams. Published by: E. Stock, 1906.

Page 78 (top): Zee Prime, CC BY-SA 2.5

Page 78 (bottom): Classical Numismatic Group, Inc.

Page 83: Bartolomé Esteban Murillo

Page 84: Photo courtesy of Dr. Elmar Buchner

Page 85 (top): Adli Wahid, CC BY-SA 4.0

Page 85 (bottom): saudipics, CC BY-SA 4.0

Page 91: Ludovisi Collection, Museo Nazionale Romano

Page 94: Sir Godfrey Kneller (1689)

Page 97 (top): "Diebold Schilling-Chronik 1513," Eigentum Korporation Luzern

Page 97 (bottom): Daderot, CC0

Page 101 (top): Public domain

Page 101 (bottom): Eunostos, CC BY-SA 4.0

Page 103: Public domain

Page 108 (top left): Mike Thornton, March 2003

Page 108 (top right): Chemical Engineer, CC BY-SA 3.0

Page 108 (bottom): John Russell, 1745–1806

Page 110 (top): King, Edward, 1735?–1807

Page 110 (bottom right): Steven Klein, GQ © Conde Nast

Page 115 (top): Public domain

Page 115 (bottom): Adolf Vollmy

Page 142 (top): Public domain

Page 142 (bottom): Henrique Alvim Corrêa

Page 147: NASA

Page 150 (top): NASA

Page 150 (bottom): J. William Schopf, "Life on Mars: Tempest in a Teapot? A First-Hand Account," Proceedings of the American Philosophical Society, Volume 143, Number 3, pg. 373 (1999)

Page 153 (top): Palauenco5, CC BY-SA 4.0

Page 153 (bottom): Provided by Ian Hunter

Page 155: NASA

Page 167: Public domain

Page 179 (both photos): Antarctic Search for Meteorites Program, Case Western Reserve University/Emilie Dunham

Page 181: American Museum of Natural History

Page 189: University of Alabama Museums, Tuscaloosa, Alabama

Page 191: Vokrug Sveta, 1931

Page 194: Alex Alishevskikh, CC BY-SA 2.0

Page 195: Pospel A, CC BY-SA 3.0

Page 203: European Fireball Network, CC BY-SA 3.0

Page 241: Vahe Martirosyan, CC BY-SA 2.0

Page 249 (bottom right): Courtesy Emilie Dunham, sample collected by ANSMET

Page 253 (left): Photo courtesy of Philipp Heck

Page 253 (right): Image provided courtesy of Thomas Zega

Index